International Geographical Research:
A Survey

International Geographical Research: A Survey

DR. K. L. NARASIMHA MURTY

GYAN

First Publication in 2018

Published by

Gyan Publishing House
5, Ansari Road
Daryaganj, New Delhi-110002
Phone: 9811692060
E-mail: books@gyanbooks.com

Printed at: G. Print Process, Delhi.

Cataloging in Publication Data—DK
Courtesy: D.K. Agencies (P) Ltd. <docinfo@dkagencies.com>

Narasimha Murthy, K. L. (Kalluri Lakshmi), 1949- **author.**
International geographical research : a survey / Dr. K.L. Narasimha Murty.
pages cm
Includes bibliographical references.
ISBN 9788121214322 (hardback)

1. Geography—Research. I. Title.

LCC G73.N37 2018 | DDC 910.72 23

Dedicated to

My Mother
Smt (Late) Kalluri Pallalu Jagadamba
and to
My Father
"Dharma Sastra Chakravarthi"
"Apara Apastambulu"
"Nityagrihotri"
Sri (Late) Kalluri Ramabhadra Chayanulu
and
Acharya Siromani
Sri (Late) Sripada Jagannadham
B.A, B.Sc, B.Ed
Science Assistant
K.S. R. Z.P. High School
Vegeswarapuram
West Godavary District

Contents

Foreword

I have great pleasure in writing the foreword to the "International Geographical Research" by Dr. K.L. Narasimha Murthy. Indeed I am much grateful to the author. It is no exaggeration to say that K.L. Narasimha Murthy is an accomplished author and this new venture had sustained his quest for excellence and has certainly added a shining feather to his cap.

He has dealt with the Physical, Economic, Urban, Rural, Regional Methodology of Geography, Micro level Planning & Integrated Area Development Planning and evaluation and analysis of International Publication of Research papers.

The volume is divided in to Nine Chapters. Each chapter is a landmark in the International Geographical Research. Chapter I deals with introduction.

Second, Third and Fourth Chapters deals with Physical Geography, Economic Geography and Urban Geography respectively.

Fifth Chapter deals with Rural Geography and Sixth Chapter deals with applied Bibliometric Methodology for analysis. Seventh Chapter deals with the progress of Indian Geography and Indian's Contribution to International Geography and so on.

In Eighth chapter, Dr. Murthy has made an intensive attempt to present micro level planning and integrated area development studies at District level and further micro level units and conclusions are presented in the Ninth Chapter. At the end, two research papers have been presented for beginners.

I am sure this encyclopaedic compendium on the International Geographic Research will serve its useful purpose and will be well received by researchers, Planners, Teachers and Students of Geography as well economics and all disciplines.

I take this opportunity to congratulate Dr. K.L. Narasimha Murthy on his laudable achievement well as his service to the cause of Geographical Research.

Professor S. Ramakrishna
Department of Geography
University College of Science
Osmania University,
Hyderabad – 500 007

Preface

This important volume brings together the contributions of the authors to a tune of Nine Chapters in the field for "International Geographical Research" (IGR), and it has been spread over five years both intensively and extensively dealt with the themes of environmental geography, i.e., particularly physical environment, economic environment, social environment and cultural environment having a bearing on the active, reactive and interactive elements. The present volume is specially designed and prepared to suit the needs of the students and research scholars not only to the beginners of research but also for the seasoned specialists spatially and globally in a reasonable way. The entire work encompasses to USA, Africa, Europe, Asia and Australia and the current geographical literature may prove to be quite useful and resourceful document as a reference work and the author certainly do not claim, it is one and only one volume to match the needs of thc 21st Century students and this book may indeed give momentum to early start their research activity spatially.

In the first chapter, introduction has been presented 2nd, 3rd and 4th chapters are dealt with physical geography, economic geography and urban geography. 5th chapter deals with rural geography and regional geography. In the 6th chapter, applied bibliometric methodology an evaluation and analysis of international publication of research papers have been presented. Progress in Indian geography (the outcome of 33rd International Geographical Congress) by Indian geographers has been presented in the 7th chapter, where as Micro level planning and integrated area development themes are presented in the 8th chapter.

Finally conclusions are presented in the 9th chapter and the material relates to the global structure, besides international status in nature. Two research papers have been presented To enable the bigners to identify and recognize the other patterns of research papers and also to understand the changing patterns of resource base and pattern of development particularly at the village administrative unit in a mndal besides the application of technology remote sensing and G/S in modern geographical science.

Hyderabad **Dr. K.L. Narasimha Murthy**

31 December, 2017 Author

Acknowledgements

The author is highly indebted to Professor P.V. Sharma (R), MA, M.Sc, Ph.D, Department of Economics, Andhra University, Vishakapatnam and further indebted to Professor R.B. Singh, Vice-President, International Geographical Union (IGU), Department of Geography, Delhi School of Economics, University of Delhi, Delhi-110007. Professor Kalpana Markendeya, Department of Geography, Osmania University, Hyderabad has provided important base material to the author and the author is highly indebted to her for the timely help.

Professor S. Rama Krishna(R), Department of Geography, Osmania University, Hyderabad has extended the academic help as and when the author consulted him, thanks for the same. Professor P.S. Tiwari (R), University, of Madras, Chennai, Tamilnadu, is indebted to him and he is a source of courage and encouragement for this academic activity besides Professor N. Nagabhushanam (R), S.V. University, Tirupati and Professor M. Sambasiva Rao (R), Sri Krishna Devaraya University, Anantapuram, Andhra Pradesh and Professor M. Hari Krishna, Dept. of Geography Andra University, Vishakhapatnam and no words will reflect the heartfelt gratitude to Chy. K. Ramakrishna Sai Prasad, USA for computer online activity and Sri K.C.S. Murthy, Hyderabad to the timely encouragement for academic upgradation from time to time. The author expresses the heartfelt gratitude to various persons and institutions without whose help this volume of work would not have been successfully completed in a volume form. Further, the author expresses his gratitude to his wife Smt. K.V.S. Lakshmi, daughter, K.V. Jagadamba, Son Chy.K.V. Sriram, Chy. T. Madhav son-in-law and Two Kids, Chy.T. Surya Manasa and Chy. T. Phaneendra Aditya created the normal environment to complete this Venture. Finally, the author thank Gyan Publishing House, Delhi for their nice printing of the volume in time.

Chapter - I

Introduction

"Geo" means 'the earth' and 'Graphien' means "the description" that is to state that the subject of geography has been derived from the two words which would combinedly give the essence and essential flavor of understanding the physical, economic, social and cultural life of human being on the earth over a period of time, space & time study. It is further to state that, "Geography" is the study of the assemblage of physical, economic, social, cultural and regional aspects of the existing current phenomena upon the earth and it would be visible well from the action, reaction and interaction by the men within a specified and identified environment. It is being reflected in its time form, spirit and right perspective to prepare the current proposed document and it should be passed on to the generation as a learned work as, "International Geographical Research (2017). The nature has nurtured the physical environment into these stages and the same is compared to that of a river cycle i.e. Young, mature and old stage and have been identified into structure, process and stage. It would be seen by human eye and beyond this physical level, many physical, chemical and bioprocesses are largely or partly responsible to allow further changes either in its physical form such as the structure, the process and the stage or cultural form of the same order to visualize this current phenomena of the ongoing process, particularly geographer's role at the International/National/State and District level in the form of valuable contributions to the subject and to the people which would be very much and highly rests upon the demand side in analysing the above processes by presenting the phenomena of changes over a period of time, space & time analysis contributes to the progress of geography of global level and its changes may be evaluated in this volume, "International Geographical Research" for people to understand the process, which is an urgent need of the geographer to evaluate the current scenario".

Major branches of geography are as follows:

(i) Physical Geography

(ii) Economic Geography

(iii) Social Processes - Space, Time and people (STP methodology) - Their action, reaction and interaction processes.

(iv) Urban resources and processes at work particularly space - Time and people.

(v) Regional resources - An evaluation.

(vi) Micro level-unit, as a source and resource for macro development processes for progress of current geographical research trends from America to Anantapur. A word about the world's geographical research trends and development paradigms and patterns are presented in this work .

(vii) Role of Technology - R.S, GIS and GPS in geographical research.

In the present research frame, one to seven items will be covered as core chapters to identify and recognize current trends in international geographical research besides introduction and conclusion as tentative chapters. Finally, the latest branch of geography is enriched into the available analytical techniques in geography which are the quantitative techniques such as R.S, G.I.S. and G.P.S. for assessing the development of geographic phenomena and there is a felt need to examine, adjudicate and adjudge the process of geography as development of geography over space and time. The text of the work deals with (i) Need to conduct a survey, (ii) Data Collection at the international level, (iii) data Collection at the state's level and district level as a platform, (iv) Big Gap, (v) Bridging the gap, (vi) Individual efforts of the author, (vii) Spatial units for sampling, (viii) Objectives, (ix) Time element, (x) Tentative chapterization of the report, (xi) conclusions are dealt with. This work has been thus titled as "International Geographical Research: A Survey (2018).

1. Need to conduct a Survey

Earlier work: The earlier contributions by geographers to the subject of geography was descriptive, qualitative, and analytical in its form. It was not properly evaluated by any identified agency. The premier work which was conducted and brought-out by only one organization through its journals internationally. There was no single organization with the help of reputed professors in Geography of the international level and has not been evaluated and the output of intelegentia in the report form in a book so far. The felt need has been/identified by the present author and the attempts are being put in that direction and wish to bring-out to present voluminous work in a report form and the same is entitled as "International Geographical Research : A Survey 2018. It would be considered as an important document and an asset of Geographers of the international, national, state and district levels and the sub-regional themes would be thoroughly examined and considered to assess the work for an assessment. Fortunately, in India the reputed international Professors in Geography have brought-out in series the titles continuously in the form of enriched reports. They are; (i) Survey of Research in Geography (1972, 1975, 1977, Progress of Geography in 1980 and a survey of research in physical Geography 1987 and so on. Five Year Reports were also prepared to highlight the progress of research in order to highlight the progress of research in geography in India, wherein the subject of geography has been ever changing at the global level because of its dynamic structure and were downloaded and up-loaded the knowledge into various titles to cater to the needs of the society as well to solve some of the problems which are surfaced and centered as physical, economic, social and structural forms and also to find amicable and amenable solutions for the aforesaid problems through development plans in India.

Keeping in view of the present international scenario in mind, there is a dire and essential need to evaluate the development of International Geographical Research Trends: A Survey with an intensive effort over an extensive geographic area representing the Continents upon the globe that have been speared-over. They are: (1) North America, (2) South America, (3) Africa, (4) /- Asia, (5) Australia and (6) Europe in order to create the database and also to bring-out a volume entitled "International Geographical Research: A Survey (2018) both qualitatively and quantitatively and prepare a status report at the International level besides the National/State and district levels in an hierarchical order such as International Geographical Research, National geographical research/state level geographical research and district level potential work in order to identify the development of geographical knowledge through teaching and research from America to Anantapur (Americas -Africa-Australia, Europe and Asia - India- Andhra Pradesh - Anantapur district as a case study in order to create data base and models in geography. It is to state that from micro level to macro level development of research in geography and global scenario could be visible in its current trends too.

2. Data Collection at the International Level

In the present research, the author has conducted an in-depth enquiry and a survey in order to recognize and identify the gaps in the existing literature, covering both the covered and uncovered aspects in various forms, such as published material and unpublished material relating to international geographical research trends from the following journals as a learned representative/referred work (material). They are:

International Platform:

1. Annals of the Association of American Geographers (AAAG), Journal, Washington, D.C, U.S.A.
2. Canadian Geographer, Canada.
3. African Geographer
4. Scottish Journal of Geography
5. Geographic Annaler
6. Australian Geographer

Asia - India:

1. Annals of the National Association of Geographers, India, New Delhi
2. The Geographer, Aligarh
3. Transactions - Pune
4. Indian Journal of Regional Sciences, Kolkata (formerly housed @Indian Institute of Technology, Kharghpur, West Bengal, India)
5. I and 2 are from one and only state - Uttar Pradesh, Bowl of Geographers of North India.

State Level (Group & Individual State Journals) were also referred to and the material has been presented in the text of surveyed potential in the chapters exclusively.

Conference Material of the International - Indian Geography Congress seminars were conducted at different geographical centres such as University centers and their contributions to geographical knowledge is also an important addition for assessment of the progress of geographical research.

District Level Platform

Anantapuram - S.K. University - Anantapur, Andhra Pradesh State is an important centre and is an important centre in Andhra Pradesh State but individual contributions outweighs the material form contains more than a geographical journal and its current performance has been assessed and presented in the7th Chapter.

3. Big Gap

To visualize the development/progress of geography, which has been periodically evaluated by the subject experts from the available literature and the same has been brought-out in volumes to cater to the needs of the researchers over a time tag 2018. Here in, a wide gap is visible, felt and has been identified, are some of the important aspects in the development of geography i.e. at the International level, the development of geography indicating its vertical and horizontal progress is clearly missing, which presents as an indelible mark of gap and to fill in the same gap, it is apt by the geographers to undertake such a need and demand based work and which has to be periodically evaluated and tested by the subject experts and the potential has been presented in a volume form, entitled as, International Geographical Research and the Current Research Trends (IGR&CRT) have been prepared by the author.

It is planned in order to bring-out an international/national / state/district level development of geography, particularly the current research activities/ trends have been identified, recognized, collected, collated and finally a valuable work was brought-out into volume, which helps the beginner researcher in geography as well as for evaluation by the international/national assessment and accreditation is done, based on this work of development. Finally, duplication of work should be stopped/arrested totally and an innovative and knowledgeable work may be undertaken by the young geographers for pursuing M.Phil / Ph.D. Courses in future both by the Indian and foreign Universities.

4. Bridging the Gap

The researcher has made an indepth attempt to find the identified gaps, to fillin gaps and the author has conducted an inquiry, a survey and equally it is also the primary work of geographers at the International/national level for evaluation. They have to develop the analytical techniques for identification of the space (s) element for measuring the area differentiation. The quantitative techniques such as remote sensing and Geographic Information System (GIS) are the tools of geographers at present for creation of spatial data base and analysis through mapping techniques, which helps to effective decision support system, are some of the modern developments in geography. The above aspects are being reflected in the research as Trend Report (TR) and development activities. The progress is slow but promising as a result of the changes and the same has been brought out & may be utilized in an academic programme.

5. Individual Efforts of the Author

Five Volumes were published and presented by the author to assess the development of geography in India which would be considered as a base for evaluating the present work of the International level.

1. Research in Geography: A Survey (1992) has been prepared by the author and was published by Ashish Publishing Company, New Delhi.
2. Geographical Research (1999) has been prepared by the author and the same was published by the Concept Publishing Company, New Delhi.
3. Indian Social Science Research: A Survey (2012) was prepared by the author and two eminent people of Geography department, Osmania University, Hyderabad were co-authored and the same was published by Teja Publishers, Hyderabad.
4. Research Methodology in Geography: A Text Book 2014 written by the author solely which was published by concept publishing company, New Delhi and
5. Indian Geographical Research: A Survey (2017) was published by R.K. Books Pvt. Ltd, New Delhi.

The above five works form as a base and also in the line to help the current research trend of activity and caters to the needs of the stud y in full form for the evaluation of international Geographical Research: A Survey which was presented as a work has been an exhaustive report at the international level.

7. Spatial Units for Sampling

The following Spatial Units (S.Us) are selected for identifying and assessing the changing nature of geography at the International, National, State and District levels. The detailed methodology adopted in this study are as follows:

1. I Level: The development of geographical science at the international level with methodological perspectives. For example (l) North America, (2) South America, (3) Africa, (4) Europe and (5) Australia, and (6) Asia.
2. N Level: National level methodology deals with the development of geography - National Geographical Research of India A Case Study and present an Uniqueness in the spread of

geographic education and geographical research in India, i.e. from New Delhi, Aligarh, Pune and Kolkata, India.

3. D Level: Development at the state level could be evaluated as there's no university offering geography/geo-informatics at the state level head quarters i.e., capital city Amaravati of Andhra Pradesh.

4. I Level: Information of the district level methodology could be evaluated in Anantapur district with the head quarters as Anantapur and it is a growth centre for geographical education and research since 1985 was selected for a case study.

5. A Level: Aggregation, disaggregation and dissemination of Geographic Knowledge in one sub-continent, from Asia India is selected. This methodology is named as "INDIA" for evaluating the progress of geography@ the district level without any journal or publication but it is growing as "Growth Centre" to promote geographic knowledge.

8. Objectives

1. To evaluate the seminal content of work that has been turned out by the members of the geography departments @international level,
2. To evaluate the progress/development of geography in India and in Andhra Pradesh state and an endeavour to do the needful to pass on present and represent the knowledge as a guide for the future generations of the development of geography at the state level and also further,
3. To evaluate the progress of geography of the sub-state level (district) which will help the young researchers to identify the gaps in literature and also to suggest the steps that are to be taken up for the development of geography.

Keeping in view of the objectives in mind, for evaluating the above process, the geography departments are selected as centres of higher learning in geography in order to identify and recognize the progress and the scaling methodology has been adopted (Spatial Time Analysis).

Department of Geography, S.K. University, Anantapur which is a district level head quarters and turned to be a growth centre that has been offering M.Sc (Geography) since 1985 which is a micro level unit termed as I in methodology for assessment.

9. Time Element

To evaluate the development of geography, the researcher has taken 18 months period of time for undertaking the research work for 2014, 2015 and 2016 years so as to acquire and prepare the necessary qualitative data base in order to create the innovative ideas and analyse them over a period of time. Keeping in view of the academic programmes of the ongoing processes of geography and its development as well as research activity relating to the geography department leading for the award of Ph.D/M.Phil degree in geography during the period under review have been identified and presented in a report format / publishable work.

10. Report

Development of Geography over a period of time at the International/ National/ State and District levels would be revealed through this report with identified and potential areas of research. The contributors at the International and National level, who are responsible for the progress of geography would be taken into consideration and also to suggest the measures and precautions that are to be taken for the development of current progress of geography and the innovative work by the new researchers in this discipline with a multi - level, multidimensional and inter and intra disciplinary processes would be estimated and suggested for the well being of this branch of geography into multi-inter-disciplinary areas are the need of current research activity and its interactive processes.

11. Conclusion

It is concluded that this multidimensional and the multi inter-disciplinary research report will help certainly for the upgradation of the subject geography as well and various development activities could be planned, designed and executed at various regional centres whether they are International/National/State level centres so as to create a new thought not only to the outsider (Observer) but also to the internal personnel to plan and augment/start/initiate academic programmes, impact and the innovative topics in the syllabi as well to conduct the courses on research activities for achieving the highest mark both in teaching and research at various area levels such as International/National/State and SubState (District) level and the related/inter-related areas particularly in the field of planning and development or progress of the subject geography, particularly the social relevance of geography applicable to the society for its upgradation and updating should be considered at large besides the other branches of economic and physical geography for the development of society over a period of time (Space-Time Analysis).

Chapter - 2

Physical Geography: A Survey (International Level)

According to W.G. Moore, ecology is the science which treats of organisms in relation to their environment; it is frequently sub-divided into human ecology, animal ecology, plant ecology and bio-ecology, the last name dealing with the interrelationships between animal life and plant life. Ecology lies on the frontiers of so many other suggests, including various branches of Geography, that its limits have not yet been precisely defined (W.G. Moor 'A Dictionary of Geography", p. 74), Premier Reference Book, Reprinted 1971).

A.M. Macdonald defines ecology is the study of plants or of animals or of peoples and institutions in relation to environment and further identifies environment as a surrounding conditions influencing development phenomena or growth.

The word 'develop' as to exploit the natural resources of a region and 'development' as ' the act or process of developing and new situations that emerge.

Keeping in view of the above in mind 'geography' is defined as the study of physical, economic, social and cultural aspects of ecology helps for "development" (Quotation Mine Please). Geography has emerged out into six traditional branches. They are:

1. Physical Geography (Geomorphology, Climatology and Oceanography) (Land, Air, water). Prudvi (land), Apaha (water); Tejo (light) Vayuhu (Air) Akasha (Sky).

2. Economic Geography (Agriculture - Industrial- Resources)
3. Social Geography (Urban - Rural (Spatial Distribution)
4. Cultural Geography (Cultural: Spatial Distribution)
5. Regional Geography (Resources - Economy - Spatial Distribution)
6. Human Geography (Spatial Distribution of Human Societies)

The author has attempted to present the status and structure of the profile and progress of the ongoing development in Physical Geography in the following paragraphs.

The International geographical research progress has been presented into the following series: They are as follows: (1) Physical Geography, (2) Economic Geography, (3) Urban Geography and (4) Regional Geography

The traditional way of approach has been adopted, however, the current branches are also used at its minimum level. This text has been divided into the following. They are: (1) Terrain studies, (2) Paleo-climatological" simulations and climatic change, (3) Remote sensing, (4) Land use and land cover dynamics (land use geography), (5) Gaps in literature, (6) Suggestions for future development.

1. Quantitative Geography and Climatology

" If Gilberts 1914 paper was as so and as structure seems to think, why was it not i: adopted as a sign post to future work in geomorphology, instead of being largely for further and ignored for thirty years? The answer may be, as Strehler himself seems to imply, that geomorphology was a part of geography. Hydrologists and geologists did not direct their major interest towards such matters, or when did they followed Davis. The following included Douglas Johusdo, C.A. Cotton, N.M. Fennemean and A.K. Lobeck. Strahlar hold that they made "Splendid contributions to descriptive and regional geomorphology" and "have Drolited a sound base for studies in human geography, but they did not greatly advance the scientific study of geomorphologic process. This is not to say that there was no quantitative work in geomorphology prior to Strahler".

One immediate response to Stahlar's attack on Davis came from Quam, who wondered whether mathematical fomulae and statistical analysis might not quite a false impression of objectivity and accuracy. A "There has been a relevant attempt in certain quarters to devise a more than a ponderous sort of cant. The processes and results of rock sculpture and not usefully amendable to treatment by mathematics at higher certificate level of any", 'best' is to result from the movement, we have yet to see it, it will be time enough to incorporate it in the suspect when it has discovered or expressed something which cannot be expressed in place English. For ourselves we continue to regard W.M. Davis as the founder of our craft and regret the murimerings of dispraise heard occasionally from his native land"[1].

L.King is inclined to support Strahler

Statistical analysis is essentially the method of bulk sample and is admirable for the study of complex phenomena and processes into which enter a large number of variables. As yet few geomorphic topics provide data suited directly to statistical treatment, and methods may have to be adopted to the new field of enquiry, so that too facile results should not be expected. The net result must be however, a greater precision in geomorphic thinking.

Several geomorphologists, including Chorely, Dury, Mackay, Wolman and others, in addition to Strahlar, are using quantitative methods, and the practice seems likely to spread.

There has been little argument about the application of quantitative techniques to climatology. This branch of our subject embraces the most apparently manageable and quantifiable continuum that geographers have been concerned to study. Thornthwaite and Mather, Hare, Bryson and others have been applying quantitative techniques to climatic problems for some time, and with great effect. The quality of their work has virtually silenced the potential critics.

2. Terrian Studies

Atmospheric particulate matter (PM2.5 and PM 10) can adversely affect human health and also has impacts on climate and precipitation. Much research has been done on the transport of human made five grained matter in modern times. It is still unclear, though, what controls the transport process of five grained sediment from natural sources on the Millennial scale. In this study we present Holocene basin - wide fine grained Sediment records from the Shiyang river drainage basin system in Arid China. Six Holocene Sedimentary Sequences were collected from various geomorphologic units of the drainage system. A total of 1043 sediment samples were obtained for analysis of five grained sediment, fifty eight radiocarbon dates were acquired for establishing the geochronological frames. Besides this, we synthesized the results from transient Paleotclimate simulations to understand environmental back-grounds of the Holocene. Our records and simulations indicated that fine grained sediment 2.5 un-content was relatively stable and less affected by monsoon intensities and circulations. Millennial Scale fine grained sediment 10 um content varied according to sedimentary fancies and it oJ was negatively correlated with the winter monsoon intensity at aeolian sediments. The fine grained sediment 10 um content increases dramatically at lacustraine layers in lake and alluvial sediments from the middle and lower drainage basin, showing its relationship with long term moisture conditions that are closely related to monsoon precipitation based on climate simulations. This finding contributes to our understanding of the fine grained matter trends against the backdrop of the global warning[2].

The recent availability of high-resolution digital elevation data has revealed numerous large landslide scars, in glaciomarine deposits in Western Sweden. In order to determine the timing and cause of six such landslides, geomorphic and stratigraphic investigations were undertaken. Based on cross-cutting relationships between the landslides Scarps, raised shorelines and other wave washed deposits, the geomorphology indicates that the slides occurred along early Horocene Coastlines. Parts of each slide occurred be low and above sea level. Strategraphic investigations at the 6 landslides reveal the presence of glaciomarine fine sand below glaciomarine clay and the presence of confined aquifers beneath the landslides given this stratigraphy and the geomorphic link to the coastlines, it is interpreted that the emergence of the ground water recharge zones above sea level allowed pore pressure to build within the aquifers and led to slope failures. It is suggested that the mechanism linking falling isostatic sea level to these near shore/coastal landslides may also apply to continental slope landslides during periods of falling eustatic sea lever[3].

The concept of digital Earth as a digital replica of the planet has its roots in science fiction, in the writings of environmental visionaries and in a speech prepared for delivery by Vice-President Gore in 1998. Five use cases are identified that together comprise the vision of the Digital Earth, a geoportal, a visualization service, a platform for simulation and prediction, a source of unprecedented spatial and temporal resolution and technology fully integrated into human activities. Progress to data is reviewed for each use case, along with the research that would be needed to achieve each aspect of the overall vision[4].

A major goal of landscape ecology is to understand the formation, dynamics and maintenance of spatial heterogeneity. Spatial heterogeneity is the most fundamental characteristic of all landscapes and scale multiplicity is inherent in spatial heterogeneity. Thus, multi-scale analysis is imperative for understanding the structure, function and dynamics of landscapes.

Although a number of methods have been used for multiscales analysis in landscape analysis ecology, since the 1980's, the effectiveness of many of them, including some commonly used ones is not clear questionable. In this paper, the authors have discussed the approaches to multiscale analysis of landscape heterogeneity of the direct and indirect approaches. They focused on scale variance and semi-variance methods in the first application approach and 17 landscape matrices in the second. The results show that scale variance is potentially a powerful method to delete and describe multiple scale structures of landscapes, while semi-variance analysis may often fail to do so especially if landscape variability is dominant at broad scales over five scales. Land scale matrices respond to changing grain size rather differently and these changes are reflective of the mollifiable areal unit problem as well as multiple scale structures in landscape pattern. Some matrices (eg., the no.of patches patch density, total edge, edge density, mean patch size, patch size co-efficient of variation) exhibit consistent, predictable patterns over a wide range of grain size, where as others (eg. Patch diversity, contagion, landscape fractal dimension) have non-linear response curves. The two approaches to multiple scale analysis are implementing and their pros and Cons still need to be further investigated systernatically"[5].

Regional scale built-up areas and Surface Urban Heat Islands (SUH!) are important for Urban Planning and Policy formation. Owing to coarse spatial resolution (1000 m), it is difficult to use Moderate Resolution imaging Spectroradiometer (MODIS) land surface temperature (LST) products for mapping urban areas and visualization and SUHI related studies. To overcome this problem, the present study downscaled MODIS (1000m resolution derived LST to 250 m resolution to map and visualize the urban areas and identitfy the basic components of SUHI over 12 districts of Punjab, India. The results are compared through visual interpretation and statistical procedure based on similarity analysis. The increased entropy value in the downscaled LST signifies higher information content. The temperature variation with the built-up and its environs due to difference in land use and is depicted better in the downscaled LST. The SUHI intensity analysis of four cities (Ludhiana, Patiala, Monga, and Vatinda) indicates that mean temperature in Urban Built-up core is higher (38.87 °C) and compared to sub-urban (35.85 °C) and rural (32.41 °C) areas. These techniques demonstrated in this paper enhance the usage of open source wide swath MODISLST for continuous monitoring of SUHI and urban area mapping, visualization and analysis at regional scale. Such initiatives are useful for the scientific community and the decision-makers[6].

This paper presents the first decadal quantification of glacier area in the Cordillera as Carabaya and apolobamba in Peru and Botra (14000-14039's; 69° 14 70° 19'w). They have calculated the changes in the glacier are between 1975 and 2015 and the number of glacial lakes between 1985 and 2015 using land sat images. Glacial Shrinkage Change in this study region was most likely to have causal the expansion of numerous glacial lakes in recent decades. In contrast to observed glacial fluctuations in the Western Cordillera as of the tropical Andes, continuous glacier shrinkage was observed in the Western Cordilleras of Paru and Botina. The estimated glacier mean area loss between 1975 and 2015 in the Cordillera Apolobamba and the Cordillera Carboy was about 51% and 79%, respectively Glaciers Located below 5000 m.n.sl nearly disappeared (930/0 and 95..5%) in the Cordillera Apolobamba and the Cordillera, respectively during this period, but glaciers above 5000 m.rsl were relatively stable (13.5% and 31% in Apolobama and Caraway respectively). Further on glacier area loss by aspect reveal that glaciers oriented towards an Amagizan Basin, were retreating faster (74.5% and 86% respectively than those oriented towards the Pacific Coast (40% and 68%, respectively). They observed the formation of 206 new glacial lakes between 1985 and 2015 in these two mountain ranges with an increase in the total area of 3.8 krrr', which is a significant growth injust 30 years[7].

Spectrally similar nature of land covers in a glacierized terrain hampers their automated mapping from multispectral satellite data, which may be overcome by using multisource data. In this study, an artificial neural network (ANN) - based information extraction approach was applied for mapping the Kolkhoz Glacier and adjourning areas, using land sat TM (Thematic Mapper) data and several ancillary layers such as image transformation and topographic attributes. Results reveal that ANN (highest overall accuracy (O.A): 83.74%) out performs maximum likelihood classifier (highest OA:66.90%) and the incorporation of ancillary data into the classification process significance enhances the mapping accuracy (>9%), particularly the addition of near Hifrazed Red/Short Wave Infrared (NIR/SWIR) data to the spectral data. A nine-based continuation dataset (SFectral data, slope, Red/NIR and deconclation stretch) was found to be the best multisource data set. Results of the Z-tests (at 95% confidence level) also corroborate and statistically validate the above findings[8].

Climatology

Dominant components of spatial-temporal variations were derived from principal component analysis (PLA) of spatially high density data on air temperature and pressure over the Kanto region in summer, obtained from our own observation network. Their connectivity is discussed from the viewpoints of meteorological conditions and divisional variations. The spatial patterns of the three major components for the pressure and temperature fields 'results each other in view of their contribution and the temporal variations in these scores also show statistically significant correlation.

The major characteristics of the cultural variation air take-up and pressure fields on sunny days can be summarized as follows: (1) inter day time, a relatively high temperatures, low pressure area is located in the inland area and there a southerly Seabuze gradually provides over the whole Kantoplain as it develops, and (2) develop the right time after sun set, the urban heat island phenomenon in the core of the Tokyo metropolitan areas and northern suburbs appears clearly, corresponding to the low pressure area[9].

Geographic variations in air and soil temperatures are dependent on several biotic and abiotic factors "Air temperature has mostly been used to characterize thermal conditions for plant life and studies of bioclimatic gradients. From biological point of view, it is also essential to know to what extent soil temperature is coupled with air temperature. In this study, they have quantified the deviations between soil and air temperatures along gradients in latitude, altitude, and possible effects of the vegetation. Sixteen different temperature variables were estimated from 49 vegetation plots on 19 mountain summits along the high mountain range in Norway, ranging from 230 to 1780 mzsl and from 590N to 710N Soil and air temperature variables were estimated from the study plots during one year. All air and soil temperature variables were significantly completed but the rating explanation was mostly relatively low (370-60.00/0), except during the growing season. Starting the grow up seasons, determined by air or soil temperatures, could deviate by 38 days mainly due to effects of frozen soils vegetation composition, especially the Lichen cover, heel a major impact on soil temperature. Dwarf Shrub cover increased significantly with increasing July temperature. Lichen abundance and degree of soil first were strongly completed and explained a major part of the variation in soil temperatures. The study indicates that air temperature is generally a poor proxy for soil temperature in cold areas, except during July[10].

The winter of 2013-14 brought extreme storminess to the U.K. Monthly rainfall was over 200% of the 1981-2010 average. The damage was considerable, affecting landforms, habitats, societies and infrastructure.

Over that winter there were several high magnitude storms, occurring at very closely special time intervals, which were particularly devasting in the West and South West of the U.K. Universally, in the east a single extreme storm, which occurred on the December, 2013, did most of the damage. This article highlights how new technologies are being used to quantify some of the Coastal impacts of such events. The data collected support the unevenness of the impact of the storm on the shorelines at different locations around U.K. Storm had a fair leading and devastating impact along South - West-Coastlines because these Coasts are wave dominated and every storm included high magnitude worm. However, the east coast shorelines, in particular in east Angli, eve tide dominated and take a special combination of circumstances to generate the largest impacts. The storm on 5th December, 2013 generated elevated water levels in the sea, reaching even greater elevations than the notorious storm of 31 January, 1953, where lives were lost. With large ways adding to the impact of high water levels, effects were felt around the whole of the East of East Anglia, Enhanced forecasting capacity and strengthened Coastal defences meant that no lives were lost. However, future sea level rise and managing coastal storm tracks may bring even greater coastal devastation; therefore, we need to be prepared to develop new strategies for forecasting, early warming and evacuation planning[11].

The study presents the influence of climatic changes on the annual stream flow of the Nu (Salwen) River, over the period from 1958 to 2004. While the flow of the Mn Pieber is mainly influenced by precipitation, the role of temperature in governing stream flow appears to be increasing as the climate warms. The first Principal Component (PCA) analysis, extracted from climatic data accounts for its 2.50/0 of observed variable in streams flow over the period 1958-2004. A transition toward a warmer, more humid regional climate has been ongoing since the 1980's, as child could by increases in temperature, precipitation and stream flow in the Nu River Basin. A continuing trend toward higher temperatures and more intense precipitation will likely lead to increased Nu River Stream flow in the future synoptic climatology (SC) indicates a link between extremes in annual Nu River stream flow and anamolies in atmospheric circulation patterns such as the mid latitude Westerlies and the south Asian Summer Monsoon[12].

Few studies to date examine the cumulative effects of climate change and resource development on social-ecological systems, such as watersheds. In partnership with a watershed alliance we edeavoured to understand the interplay of climate changed and resource exploitation imputes in the Nechaka River Basin, which drains an area of 52,000 KM^2 in with Western

Canada as part of theFraser River Basin. Existing regional climate studies show a clear warming trend over the last century and a projected mean temperature increase of 2°C for the 2050 Watershed stakeholders determined that hydroelectric development and forestry activities interrelate particularly closely with climate impacts related to ecolysterry function, water supply and (resultantly, community well-being. Mining and Oil and Gas development do not currently represent a high level concern, but will become problematic with further resource exploitation or climatic changes. The process was effective in grounding specific, day to day realities of climate change. The results, illustrate the importance and challenge or contenders both impact sources in tandem and should inform ongoing regional planning and monitoring efforts. A major barrier to performing this type of assessment with regional stakeholders in communicating, cross referencing and visualizing the interrelationships between climate change and resource development, which are independently highly complex[13].

"In recent years, climate change has increasingly came to be seen as one of the principal threats to future global financial stability. This article identifies and critiques the emerging consensus among international financial regulators as to how this threat - the key perceived components of which are also delineated - can best be managed. It shows that the preferred approach minus hegeneoice post financial crisis regulatory practice vis-a-vis financial stability risk more generically: prioritization of market discipline under pinned by risk disclosure. The article characterizes this approach as a quintessentially neoliberal modality of governance. It also argues that in so far as this approach relying on financial market workings and financial institutional behaviors explicitly believed by the financial crisis, it risks precisely the type of "climate Musky moment" regulators aim to avoid[14].

"In this research, the spatial synoptic classification (SSC) as weather type schemes, is used as an alternative method of demonstrating evidence of climate change in the Eastern United States and Southern Canada. Changes in frequencies for the seven SSC Weather types were assessed for summer trends (May - September) at twenty eight states and also of four regions of latitude between 1950 and 2015. Using the sse, results show significant summer decreases in dry polar (D.P) demand transitional (TQ) days and significant increases in most tropical (MT) days. The North region exhibited the greatest breadth of significant results among all weather types. The D.P and TR decline was transport at higher latitudes and weakened approaching the subtropics. The MT gain was strongest across the multitudes but statistically significant in all four regions. 'The four remaining sea weather types should more localized statistically significant trends. Results suggest that these trends in weather types frequency are an indicator of summer climate change with some stations losing over 50 percent of their DP Frequency, improve 40 percent of their TR frequency and gaining over 30 percent of their MT frequency since 1950.[15]

"Recent years have seen significant academic attention to conceptualizing climatic justice and how its ideas might be mobilized in political debates on climate policy. This article highlights to these debates by advancing two arguments. The first concerns the need for greater examination of how climate justice co-exists and concepts with more established political and justice considerations during the negotiation of climate policies. argue that distinguishing analytically between .normative interpretations of climate justice and justice claims made by parties affected by climate change or by mitigation an adaption policies provides fertile ground for deepening understanding of the multivalent and relational nature of climate justice and confronting challenges to its incorporation into climate responses. The second argument concerns the importance of exploring how proponents and opponents of climate action strive to develop, "spatial anchors" for justice claims to increase their legitimacy in policy debates. Based on analysis of carbon pricing controversies in Australia, the article illustrates how supporters of carbon pricing initiatives stressed international justice issues, whereas opponents mobilized multi-scalar and multivalent international, national, regional and local justice narratives to gain traction for their arguments. It concludes by calling for further investigation of the multivalence of climate justice and of how climate justice might be specially, represented to advance its leverage in political debates on climate policy't[16].

Remote Sensing and GIS

Great strides have been made in geographic information systems and science over the past 14 years; through the development of spatial data infrastructures and the infrastructure of data sharing; through advances in the technologies of positioning data acquisition, data dissemination, and the data analysis, and through advances in the science that lies behind the technology. Five further scenarios are examined; a world in which it is possible to know where everything is at all times; a world of positioning, representation and way finding that is fully three dimensional; a world that involves the citizens as both consumer and producer of geographic information, a world of ready access to predictions of future landscapes; a world of real-time dynamic information[17].

Similarity between the objects has close association with the geometrical structure and details of the objects. Therefore, in this, a framework was presented based on the geometric criteria. The Fuzzy Membership Functions (MF) and human spatial cognition. To increase the efficiency of MF and for the problem to get closer to reality. 1K, 2K, 5K, 10K, 25K, 50K and 100K data were used in this framework. The results are thus that the degree of 8 partial similarities is specific to the objects themselves and their geometric structure. No linear or non-liner relationship could be found between scale changes and degree of spatial similarity of the objectives because the geometry of any object independently affects its degree of similarity to other objects in different scales and sources[18].

It is an observed fact that the applications of remote sensing (RS) techniques to monitor ecosystems services has increased in recent years. Nevertheless, the potential application of R.S to monitor some of ecosystem services is still challenging. The paper reviews, the application of R.S to monitor ecosystem services of forests, manpower and urban areas. Satellite data provide substantial information about dynamics of environmental changes over time from local to global scale. This information are useful data sources for the people who are involved in the ongoing evaluation and decision making process to manage ecosystem. Many recent research papers on the topic were reviewed to find new application and limitations of R.S for monitoring ecosystem services. Advanced R.S techniques have high potential to monitor ecosystem services with the advancement of sensors ranging from aerial photography to high and medium resolution optical R.S. and from hyper spectral R.S. to micro wave R.S.[19].

Spatial Thinking & Geographic Information Science

This article discusses spatial thinking in geographic information science (GI Science), through an empirical examination of experts and students geospatial conceptions and thematic map reading. The first study examined the structures in which GI Science concepts are conventionalized by experts and students. In experts' conception, clusters for geo-partial data, GIS applications, and geospatial entities - operations - relations, and maps were identified. In students conceptions, with the terms international loss closely. High-spatial students' conceptions corresponded to those of experts to a greater degree. The second study examined geospatial reasoning and showed that thematic map reading consisted of various components differing in their relationship with spatial ability. High spatial students tackled thematic map reading by identifying more spatial distributions and comparing multiple maps more frequently. They did not necessarily make more statements about reasons for the observed patterns. It is important to distinguish spatial thinking and thinking about space, the latter of which involves geographical, beyond purely spatial, components 2016 by Association of American Geographers initial submission[20].

Land-Use & Land Cover Dynamics

Based on four time individuals within a forty year period of observations we construct Land-Use/Land-Cover (LULC) maps and estimate the transition probabilities between six LULC states. The maps and Transition Probability Matrices (TPMs) were built based on the high resolution aerial photos and 30m multispectral land set images for the same years. We considered the TPM constructed from manual classification of the aerial photos as a reference and compared it to the TPM constructed from the land-sat image classified with several methods: mean shift segmentation followed by random forest classification and three pixel based methods popular in Cellular Automata (CA) studies: K-means, the iterative self-organizing data analysis techniques (ISODATA), and maximum likelihood. For each classification method, the TPMs were constructed and compared to the TPMs for the aerial photos. We prove that the goodness-of-fit of maps obtained with the three pixel based methods was insufficient for estimating the LULCTPM. The LULC maps obtained with the object-based classification fit well to those based on the aerial photos, but the estimates of the TPMs were yet qualitatively different. This article raises doubts regarding the adequacy of land-sat data and standard classification methods for establishing LULCCA model work. We appeal for a new view of the CA modeling methodology. It shall be based on a long-term series of carefully validated LULC maps that portray different types of land-use dynamics and land planning systems over long and representative periods of population and economic growth[21].

Land-Use Geography/Political Economy

Not until the beginning of the twentieth century did US City government turn to comprehensive zoning to gain control of their land-use and built-environment. Nineteenth Century Cities had comparatively unregulated land-use systems, where properties and builders found minimal restrictions to their choices to develop urban land. This article exploits newly digitized geographic information systems (GIS) data at the level of building foot prints, made availably the New York Public Library, to study the land-use geography of mid Thirteenth Century Manhattan, the Western World's then third largest city. We ask what was the spatial order of the Fourth Century City. Beyond the Case, what can we learn about land-use in political-economy where market forces operated with much greater freedom? Addressing these issues, we introduce a variety of advanced

GIS Methods to the original data set. Specifically, we examine the separation and mixing of the three basic land-use types of commerce, industry and residence, by the spatial units of both houses and streets. In addition, we measure at a new level of precision the enormous variations in residential density and crowding that defined the growing sociospatial in equalities of 19th century cities. Documenting systematically and in detect these spatial patterns, including their socially undesirable outcomes, helps understand now 19th Century Cities developed and the conditions they produced to warrant increasing land use controls, from building codes to government-mandated zoning[22].

Analysis

1. It is noted that the profile of geographical research in physical geography at the international level has also been with its deficiencies even though efficient work has been turned out. The gaps are still persisting.
2. Descriptive geography attached with cartographic skills has ended almost by 1960 and the quantitative techniques such as techniques particularly statistical methods have been used to a maximum possible extent. Many geographers were trained and they have adopted the quantitative techniques in geographical research at various area levels.

Gaps in Research

1. Geomorphology and climatology have find good place in physical geographical research and the geographic researches have not find an apt place at the international level.
2. Pale climatological studies are at the minimum level and future researches have to take-up this branch for their career development. It is pertinent to take out physical oceanography, chemical oceanography and beach studies have been developed individually, as this branch offers an M.Sc, (Tech) in Metrology and Oceanography. Keeping in view of this) many geography

departments have not putin their efforts to take up 'oceanic' studies after 1980's.

Task Ahead

1. Landform and land-use - at the national level and also at the district level.
2. Regional geomorphology - Micro level study at the National and state level.
3. Terrain studies - Terrain studies with an application of aerial photo interpretation in relation to changes over a period of time.
4. Land-use and land-cover dynamics - land-use geography topics relating to landform evolution and evaluation with quantitative techniques.

Conclusion

It is concluded that physical geographical research has been ever changing because of its latest techniques that could be used to identify the structural composition besides physical land-forms. Surveying, mapping of landforms and topo-sheet interpretation along with aerial photo interpretation besides remote sensing techniques and GIS are important methods to study in detail the physical geographic process. Besides the above, the author feels to evaluate the ongoing international geographic research in economic geography is quite important and apt for its presentation. Economic Geography: A Survey (International Levels is presented in chapter-3 in order to identify the structural process and also to suggest new fields of research in it.

Notes

1 Ian Burton, "Quantitative Geomorphology and Climatology", Canadian Geographer, VII. 4, 1963, p. 151.

2 Yuli Pengchengli, Chengqizang & Yue Wang, "Long-Term Fine-Grained Sediment Records in a Praihage System in Anil China: A New Perspective from Paleo Climatologically Records and simulation", pp.1216-1228, Published online: 2nd April, 2017, Annals of the American Association of Geographers, Washington, DC, Yol.1 07, Issue No.5, 2017.

3 Colby A.Smith, Olof Larson & Mats Engdahl, "Early Holocene Coastal landslides linked to land uplift in Western Sweden", pp. 1-24, Geographic Annular, Series A, Physical Geography, 2017.

4 Micheal F.Goodchild, "The Future of Digital Earth". Annals of GIS, Vol. 18, No. 2, 2012, pp. 93-98.

5 Jianguo Will, Dennis E. Jelinski, Matt Luck & Paul T. Tuelhar, "Multi scale Analysis of Landscape Heterogeneity: Scale Variable and Pattern Metrics", Geographic Information Sciences, A Journal of Association of Chinese Professionals in Geographic Information System, Vol. 6, Issue No. I, 2000, pp. 6-19.

6 Sandip Mukherjee, P.K. Joshi and R.D. Garg, "Analysis of Urban Built-up Areas and surface urban heat island using downscaled MODIS derived land surface temperature data", Journal of Geo-Carto, International, Vol. 32, No. 8, 2017, pp. 900-918.

7 Bijeash Kozhikkadam Veethil, Sergis Floreneio de Souza, Jefferdon Cardia, Simoes & Sebastian Fellipe Ruiz Pereira", "Decadal Evolution of Glaciers and glacial lakes in the Apolobamba-Carabaya region, tropical Andes (Boliria-pern.) Geographic Annaler, Series 'A "Physical Geography, pp. 1-14.

8 Aparana Shukla & Bisma Yousuf, "Evaluation of Multisourc e data for Glacier remain mapping: a neural net approach', Journal of Geocarto, Vol. 32, No. 5, 2017, pp. 569-587).

9 Yamato Hiroaki, Morishima Wataru, Akasalkumi, Mikami Takehiko, "Characteristics of Spatiotemporal Variations in Surface temperature and temperature fields detected from spatially High density observations over the Kanto Region in Summer, Utilizing surface Pressure and Temperature", Data from July to September,2014, EJournal, Geo, Vol. 12, No. 1, Japan, 2017.

10 Arvid Odland, Gauri Bandekar, lmger Hanssen-Baver & Sylvi M. Sandvik, "Relationships between Vegetarian, air and soil temperature or Norwegian Mountain Summit", Geographic Annalar, Series A, Physical Geography.

11 Sue Brooks, "Coastal residence and vulnerability: Storm impacts, extreme weather and regional variability in the U. K, Winter, 2013-14", Biorn-Birbrack Institutional Research on line. 2017, Volume 102, lssucs-Z, date Summer, 2017, pp. 60-70.

12 Huiqin Wang & Fengchen, "increased Stream Flow in the Nu ~r (Salween) Basin of China, due to Climatic Warming and Increased Precipitation", Geographic Ann a~er, Series A. Physical Geography, so" June, 2017, pp. I-II.

13 Ian M.Picketts, Margot W. Parkes, Stephen J. Dery, "Climate Change and Resource Development Impacts in Watersheds: Insights from the Nechako River Basin, Canada, First Published: 2 December, 2016, full publication history, Dol: 10.1111/cag.12327riew/save, citation. Canadian geographer, Vol. 61, issue 2, 2017.

14 Matthew Gandy, "Climate Change and Financial Instability: Risk Disclosure and the problematic of Neoliberal Governance", Annals of the American Association of Geographers, Washington D.C, Vol. 107, 2017, Issue No. 5.

15 Jabow C. Senkbeil, Michelle E. Saunders & Brent Taylor", Changes in Summer Weather Type Frequency in Eastern North America", pp. 1229-1245, Published online 24 April 2017, Annals of the American Association of Geographers, Washington, D.C., Vol. 107, 2017, Issue No. 5.

16 Ian Bail y, "Spatializing Climate Justice: Justic e Claim Making and Carbon Pricing Controversies in Australia", Published only 04 April, 2017, Annals of the American Association of Geographers, Vol. 107, Issue No. 5, Washington, D.C, 2017.

17 Michael F. Goodchild, "Geographic Information Systems and Science: Today and Tomorrow", Journal of Annals of Geographic Information System, Vol. 15, Issue No. 1, China, 2009, pp. 3-9.

18 Alireza Chehreghan and Rahim Ali Abbas Pour, "An Assessment of Spatial Similarity Degree between Poly lines on Multi Scale, Multisource maps". Journal of Gcocarto, Vol. 32, No. 5, 2017, pp. 471-487.

19 Ram Avter, Pankaj Kumar, Akiko Oono, Chitresh Saraswat, Singay Dovji and Zarchi Hlaing, "Potential Application of Remote Sensing in Monitoring Ecosystem Services of Forests, Managers and Urban Area", Journal of Geocarto, International, Vol. 32, No. 8, 2017, pp. 874-885.

20 Toru Ishikawa, "Spatial Thinking in Geographic Information Scienc e: Students Geospatial Conceptions. Map based Reasoning and Spatia! Visualization Ability", Article: October, 2015, Annuals of the Association of American Geographers, Washington D.C., 2015.

21 Yykua Brinblet, Michael Gilichinsky & Itzhak Benenson, "Cellular Automata Modeling of Land Use / Land Cover Dynamics: Questioning the Reliability of Data Sources and Classification Models", Published online: 15 September, 2016, Annals of the American Association of Geographers, Vol. 106, Issue No. 6, Washington, DC, 2016.

22 Grcgely Baics & Leah Meisterlin, "Zoning Before Zoning: Land Use and Density in Mid-Nineteenth Century, New York City", Published online: 26 May, 2016, Annals of the American Association of Geographers, Washington, D.C., May, 2016, Issue No. 5.

Chapter - 3

Economic Geography: A Survey (International Level)

According to A.M. Macdonald 'economy' is defined as the management of a household, administration of the material resources of an individual, community or country and the state of these resources, organized system and regular operations as of nature.

"Economic Geography is the study of economic activity at an administrative level, International, national, state or sub-state by man upon the earth to eke out livelihood" (Quotation mine please).

"The economic activities that have been designed and adopted in order to harness natural resources both traditionally and technologically" are important (Quotation Mine please) and Macro, Meso and Micro level activities which helps for earning livelihood by men/community over a period of time up-on the globe".

"Keeping in view of the above situation, economic Geography is defined as the study of economic activities in relation to environment with human skills of men "over a period of time" and the process is presented.

The entire text of the survey has been designed and organized in the following subheads. They are: (l) Economic, Geographic situation, (2) Ecology. (3) Vegetation, (4) Livestock, (5) Human Environment-interactions and sustainable development, (6) Land-use change, (7) Economy, (8) Remote sensing, (9) Geopolitics, (10) Gaps in literature, and (11) Future areas.

1. Economic - Geographic Situation

Physicists working on a micro cosmic level encounter of the same kinds of problems with quanta and energy that social scientists do with people. Social Science needs to acquire demonstrable value as a predictive science, without a corresponding need to control, restrict or requirement of the individual. A social science which recognizes random behaviour at the predictable order at the macro cosmic level is a logical outgrowth of qualitative revolution.

The observation and description of regularities, such as these in the spatial arrangement of cultural features, human activities or physical variables, are first steps in the development of theory. Theory provides the science through which myriads of facts are sorted, and without it the facts remain a meaningless jumble. Theory provides the measure against which exceptional and unusual events can be recognized. In a world without theory there are no exceptions; everything is unique. This is why theory is so important. As Braiterwaite puts it, "The function of a science is to establish general laws covering the behaviour of empirical events as objects with which the science in question is concerned to enable us to correct together our knowledge of the separately known events and to make reliable predictions of events yet unknown.

Ian Burton concludes that quantitative techniques are a most appropriate method for the development of theory in geography and any branch of geography claiming to be scientific has need for the development of theory and any branch of geography that has need for theory has need for qualitative techniques[1].

It is appropriate to speak of central place then as one relatively well developed branch of theoretical economic geography. Wolman comments that, "the emphasis on principles that Scheideggar stresses directs attention to interrelationships and hopefully lessons the tendency to observe, measure and record everything because it's there. This remark can be applied with equal value to the development of theory in other branches of geography".

Central place theory is in operation in some schools particulary in economic theory. One role of an economic geographer is to define and adopt available economic theory. In doing so, he will improve the theory he borrows.

Further concludes that Graph theory is a branch of mathematics concerned with networks and may be adopted to fit all manner of collection, distribution and communications systems. It is conceivable that a body of useful theory could be built up around the application of graph theory to geographical problems. This is an example of what is meant by theoretical geography. It is a direction that an increasing number of geographers are likely to follow. Let us hope that the effort will meet success.

Ecology

The understanding influence of multiple factors variations on Land Surface Temperature (LST) remains elusive. LST was retrieved by the atmospheric connection algorithms. Based on the correlation coefficient, stepwise regression analysis was developed to examine how multiple factors variability led to LST variations. The differences in LST between impact factors vary depending on time in a day. The elevation and land-use types significantly affect the LST in sunny slope or shadow areas has a significantly quadratic curve correlation or a negative linear correlation with 'it, the 35/ influence of slope and aspects is not very significant. LST for forest-land, grassland and bare-land in the sunny slope and shadow area was the cubic polynomial related to its elevation. Normalized difference vegetation index (NDVI) and normalized difference moisture index (NDMI) effectively express LST in mountainous. LST and NDMI or NDVI have a significantly negative correlation, NDMI is more effective and more applicable for the expression of LST[2].

Through a review of recent writings in political ecology and agrarian studies, this paper appraises the potential for emerging forms of 'green economy' initiatives to catalyze new forms of internal displacement in West Africa, with specific emphasis on the post war catalysts Liberia and Sierva Leone. Following the international committee of the Red Cross Guiding Principles of Internal Displacement, the primary drivers of forced migration are conventionally thought to include some combination of civil war, ostensibly 'ethnic' strife or cleansing, development - induced migration or allegedly 'natural' disasters . However, land acquisition for both conservation and commercial agriculture seem poised to constitute an important additional driver of internal displacement, threatening to map on to the unresolved legacies of these previous instances of forced migration in both Liberia and Sierva Leone. Drawing lessons from the role of agrarian grievances in each country's history of civil conflict, in particular, I explore the ways in which such farms of displacement may pose clear and present dangers to the very peace building efforts that propionates of a green economy transition in the Mano River region Claim to support[3].

Vegetation

The traditional methods of forest classification, based on the interpretation of aerial photographs and processing of multi-spectral and or hyper-spectral remote sensing data are limited in their ability to capture the structural complexity of the forests compared with analysis of air borne LTDAR (light deviation and ranging data. This is because of LIDAR's penetration of forest canopies such that detailed and three dimensional forest structure descriptions can be derived this study applied airborne LIDAR data for the classification of cool temperature rain-forest and adjacent forests in the Strzelecki Ranges, Victoria, Australia. Using normalized LIDAR point data, the forest vertical structure was stratified into three layers. Variables characterizing the height distribution and density of forest components were derived from LIDAR data within each of these layers. The statistical analyses, which included one-way analysis variable with post hoc test, identified effective variables of forest type classifications. The results showed that using linear discriminant analysis, an overall classification accuracy of 91.4% (as verified by the cross-validation was activated in the study area[4].

An automatic fractional vegetation cover (FVC) estimation method based on image characteristics in an agricultural region was proposed in this study to remove the empiricism in determining the key parameters of empirical methods. This method automatically determined the soil and vegetation lines in the ten-dimensional space of the red and blue band reflectance's, which involved an iterative soil and vegetation pixels selection procedure, and then estimated FVC of a pixel based on its distances from the soil and vegetation lines. The accuracy assessment using field survey data indicated the performance of the proposed method (R2=0.69, RMSE=0.072, Bias=0.014) was comparable with several commonly used empirical methods. Therefore, it was indicated that the proposed method could effectively estimate Fye in the corn-dominated region".

Human Mangrove interactions

"Unlike the global trend, the area of mangrove forest increased in the estuaries of low cosamance and Sire-Saloum, Senegal, between 1986 and 2006. We collected multisource data (social & spatial and applied a mix of qualitative and quantative analytical methods to investigate the human-mangrove interaction during this period and to understand the cause of the observed increase. This research demonstrates that after several decades of drought, the weather conditions of recent years were the main detrainment for the increase in mangrove area. Results, however, suggest that the increase in mangrove forest is not perse an indicator of sustainability and that the increase likely masked a decline in the capacity of these mangrove forests to provide key ecosystem goods and services. The surveyed committees clearly perceived a diverging trend between the increase in area and the decline in the productivity of mangrove forests. The increasing and unregulated pressure on mangrove based fisheries and unsustainable practices such as the extensive use of mangroves wood for fish smoking heavily contributed to the perception of the qualitative degradation in these mangroves. As local livelihoods are intimately linked to the productivity of these mangrove ecosystems, policy interventions should integrate the social and environmental components of mangroves to ensure the sustainability of these human mangrove coupled systems[6].

Livestock

The extreme weather events in Mongolia of deep snow; Severe other conditions that render for age unavailable or inaccessible, which in turn results in extensive livestock deaths. Mongolia is economically vulnerable to extreme events due to an increase in non-professional herders and the livestock population, brought about by a de-regularized industry. Thus it is hugely informative to try to understand the spatial and temporal trends in livestock population change. To this end, annual livestock census data are exploited and a geographically weighted principal component analysis (GWPCA) is applied to goat data recorded from 1990 to 2012 in 341 regions. This, application of GWPCA to temporal data is novel and is able to account for both temporal and spatial patterns in goat population change. Furthermore, this methodology is extended to simultaneously optimize the number of components to retain and the Kernal bandwidth. In doing so this study not only addresses the GWPCA method but provides a useful insight into the spatio temporal variations of the Mongolia goat population.[7]

Human Environment interaction

In value chain scholarship, chain governance is the relationship of power among firms in a production network. For economic geographers working on the environment, governance refers pricisely to state and non-state-based institutional and regulatory arrangements shaping human-environment interactions. Yet the theoretical and empirical lines between these two concepts of governance are opaque. Drawing on a longitudinal case study of the canned tuna value chain and historic materialist method, we demonstrate how inter-firm strategies over the appropriation of value and distribution of costs and risks work through the environment. We document moments of change in the value chain that enliven a dynamic understanding of how a lead firm becomes and reproduces its power and strategies that subordinate firms deploy to try to counter the power of lead firms. We posit that these moves broaden value chain scholarship's focus from governance typologies toward the gravitational tendencies of capitalist completion and that such tendencies are inextricable from the environmental conditions of production through which they are made possible. This approach enables us to look at value chains and the environmental conditions of production as mutually constitutive, helping to explain vexing modern environmental problems as a core element of the general tendencies, mechanisms and drivers of power in chains[8].

Sustainable Development

The authors have examined gender equality and women's empowerment (GEWE) as reflected in Millennium Development Goals (MDGs) as well as the conceptualization and localization of the sustainable development goals (SDGS). The paper adopts Amartya Sen's Women agency conceptual framework to advance the thesis of this article that while there is general consensus on achieving gender equality and women's empowerment, the MDGs adopted in 2001 fell far short of fulfilling this consensus and that the incoming SDGs must respond to fundamental issues of GEWE from both quantitative & qualitative perspectives. This paper addresses the issues of conceptualization, context, and design of MDGs and how lessons learnt in its implementation inform SDGs. The question is, will GEWE maintain any pride of place? Analysis in the paper reveals that qualitative indicators such as human rights, equity and capabilities were silent in the MDG design, constructs and implementation. The paper suggests that the understanding and putting in place of mechanisms of continuous assessment of the human - centered indicators, such as quality of life, as part of the SDGs localization efforts portend increasing effectiveness of SDGs towards addressing gender - sensitive issues of social exclusion, inequality and resource distribution, which have a great bearing on sustainable development[9].

Land use change

"Global land-use and Land-cover change (LUCC) data are crucial formulating a wide range of environmental conditions. So far, access to high resolution LUCC products at a global scale for public use is difficult because of data and technical issues. This article presents a future land-use simulation (FLUS) system to simulate global LUCC in relation to human environment interaction, which is built and verified by using remote sensing data. IMAGE has been widely used in environmental studies despite its relatively coarse spatial resolution of 30 arc min, which is about 55 k.m at the equator. Recently, an improved model has been developed to simulate global LUCC with a 5 mim resolution (about 1 km at the equator). We found that even the 10 km resolution, however, still produced major distortion in land-use patterns, leading urban land areas to be underestimated by 19.77 per cent at the global scale and global land change relating to urban growth to be under-estimated by 60 to 97 per cent, compared with the 1km resolution model proposed through this article. These distortions occurred because a large percentage of small areas of urban land was merged into other land-use classes. During land-use change simulation, a majority of small urban clusters were also lost using the IMAGE product. Responding to these deficiencies, the 1km FLUS product developed in this study is able to provide the spatial detail necessary to identify spatial heterogeneous land-use patterns at a global scale we argue that this new global land-use product has strong potential in radically reducing uncertainty in global environmental modeling[10].

Spatial and temporal evolution - Economy

"Regional Economic development is inherently Uneven as determined by the local conditions and available resources. Specialized villages (SVS) in China played a very important role in the development and economic transformation in rural areas. By integrating regional spatial structure theory, multilevel network ·theory and spatial interface theory, this article examines the spatial and temporal evolution of SVs in Henan province, China. Results from the analysis show that the development of SVS over time progressed in four stages, each corresponding to important adjustments in national agricultural policy. SVS were distributed unevenly in space and the distribution seemed to be scale dependent. At a macro level, SVS displayed a dispersed pattern over a large area. SVS showed localized clusters at a micro level, however, also exhibiting a core periphery structure. Rural economic development in China showed that SVS formed a multilevel network hierarchy we also observed that SVS were often in transitional areas including urban - rural and plain-mountain interfaces and administrative marginal zones. Finally, spatiotemporal clusters of SVS helped to identify the locations, and time period when SVS grew significantly for analysis of impacts by national policies on rural development[11].

Anglo American Economics - Economy

Over the past twenty years, a widening gulf had appeared between the increasingly internationalized financing arrangements of the World's leading corporations and the persistence of nationally compartmentalized approaches to the study of corporate control. In lieu of direct empirical evidence on corporate control at the global level, the most widespread assumption is that the globalization of ownership has turned the form of an expansion of arnis-Ienth, market based arrangements traditionally prevailing in the Anglo-American economies. Here, however, we challenge this assumption, both empirically and conceptually. Empirically we show that three-quarters of the World's 205 largest firms by sales are linked to a single global company network of concentrated (5%) ownership ties. This network has a hierarchically centralized organization, with a dominant global network core of US fund managers ringed by a more geographically diverse state capitalist periphery. Conceptually, we variegated capitalist model of contradictory market institutionalization, with the formation of the global financial marketalization. In order to understand this process of network formation, however, it is necessary to extend Polanyian model of a double movement, mediated through political interventions in the market to incorporate veblenian processes of evolutionary institutional change, mediated through the markct[12].

Global Structure

Energy markets are on important contemporary site of economic globalization. In this article, the author uses a Global Production Network (GPN) approach to examine the evolutionary dynamics of the Liquefied Natural Gas (LNG) sector and its role in an emerging global market for natural gas. They extend recent work in the relational economic geography literature on the organizational practices by which production networks are assembled and sustained over time and space, and addressed a significantly under-developed aspect of GNP research by demonstrating the implications of these practices for the territoriality of GNPS. The article introduces LNG as a techno-material reconfiguration of natural gas that enables it to be moved and sold beyond the continental limits of pipelines. They outlined the evolving scale and geographic scope of LNG trade, and introduce the network of firms, extra-economic actors and intermediaries through which LNG production, distribution and marketing are coordinated. The analysis should show LNG is evolving from a relatively simple floating pipeline model of point-to-point, bi-national flows orchestrated by producing and consuming companies and governed by long-term contracts, to a more geographic and organizationally complex production network that is constitutive of an emergent global gas market. Empirically this article gives the first systematic analysis within economic geography of the globalization of the LNG sector and it influence on global gas markets, demonstrating the potential of GPN (and related frameworks) to contribute meaningful analysis of the contemporary political economy of energy. Conceptually it pushes research on GPN to realize more fully its potential as an analysis of network territoriality by examining how the spatial configuration of GPNS emerges from the organizational structures and coordinating strategies of firms, extra economic actors and intermediaries; and by recognizing how network territoriality is constitutive of markets rather than merely responsive to them[13].

Global Change Economy

"Global changes in the economies of most developed nations have impacted the way health care is organized, even within largely public systems, and the working conditions of healthcare workers. Since the acceleration of globalization in the 1970s, service sector workers in developed nations have forced high unemployment, increased skill requirements for most jobs and a rise in non-traditional work arrangements. These sector shifts in service sector labor markets have occurred against the background of an erosion of the welfare state and growing income inequality. As well, many healthcare systems, including Canada's, were severely downsized and restructured in the 1990s, exacerbating the underlying negative secular trends in the service sector, and worse the working conditions for many health care workers. Globalization has altered the labor market and shifted working conditions in ways that have been unfvavourable to many healthcare workers[14].

Remote Sensing

"Remote Sensing, particularly, satellite imaging is widely used in scientific, government, and public applications. One of the reasons it is so highly valued is the perception of its fundamental objectivity and neutrality. Yet Iike all data satellite imagery is a product of human action elements such as specific technologies, strategic priorities, and privileged interpretations influence the availability and applications of remotely sensed data. We therefore argue that careful examination of the epistemo logical, social and political dimensions of these data is a crucial, yet relatively under-developed task, especially in the scientific literature. We conduct such an examination through the property requires (or property rights regimes are the arrangements by which rights offer particular goods are allocated, as well as the roles to which these rights are assigned and the rules that regulate this process. This framework is especially useful in revealing the large contextual variation in the production, use and appropriation of particular goods, in this case remotely sensed data. Understanding remotely sensed data through the property regimes that govern them emphasizes the political and economic dimensions of this valuable resource and reveals job embeddedness in the world it intends to capture from a far. Thus, we show that to have a better grasp on the role of remotely sensed data in science, policy and society, users must acknowledge the property regimes and other political interventions that are here shown to be indeed fundamental to their construction[15].

Digital Geography

"Everyday billions of internet users only on search engines to find information about places to make decisions about tourism, shopping and countless other economic activities. In an opaque process, search engines assemble digital content produced in a variety of locations around the world and make it available to large concepts of consumers. Although these representations of a place are increasingly important and consequential little is known about their characteristics and possible biases. Analyzing a corpus of Google search results generated for 188 capital cities, this article investigates the geographic dimension of research results, focusing on searches such as "Lagos" and "Rome" on different localized versions of the engine. This study answers these questions. To what degree is this city related information locally produced and diverse? Which countries are producing their own representations and which are represented by others? Through a new indicator of localness of search results, we identify the factors that contribute to shape this uneven digital geography, combining several development indicators. The development of the publishing industry and scientific production appears as a fairly strong predictor of localness of results. This empirical knowledge will support efforts to curb the digital divide, promoting a more inclusive, democratic information society[16].

Carbon Emissions

"Abandoned Crop Land (ACL) is often cited as land resource on 'which to produce energy crops while reducing the negative impacts of broad-scale energy crop production, for example, carbon emissions from land cover change and competition with food production. In contrast to marginal land, which refers to a set of biophysical and economic criteria usually imposed by experts or policy makers, the designation of ACL refers to a land-use decision by a landowner. As such, ACL is argued to be a more appropriate indication of land availability for dedicated energy crop production. Prevailing estimates of ACL in the United States vary widely due to inconsistent treatment of land-use conversions away from cropland and overreliance on remote sensing methods that measure land cover, even though ACL is a category of land-use. This article develops and applies a replicable and flexible methodology to estimate available abandoned crop land (AACL) at the country level in the United States, which accounts for conversion of ACL to first cover, urban development, or permanent pasture. Estimates ofAACL are derived for two scenarios, (I) Land abandoned between 1978 and 2012, which excludes lands with meaningful forest re-growth, and (2) land abandoned between 2007 and 2012, which corresponds to landless constraints imposed by the Renewable Fuel standard. Results show that 15.0 and 4.9 mhz of AACL exist in the United states in the two scenarios, respectively, amounting to between only 3 and 8 percent of total light duly gasoline consumption in the United States. The policy implications of these findings and the need for future research are discussed[17].

Reposition / Urban-Economic & Cultaral Phocesses

"What are the roles of history and memory in geopolitics? How does urban experience influence geopolitical understandings of one's place in the world? This article brings these questions to a study of Ottoman Turkish Citizens of Istanbul came to time ethnicity with nationalism and to view their Greek Orthodox neighbors as national betrayers. I propose on explicitly cultural geopolitics: an effective, embodied critical geopolitics context truly dependent on experience, encounter and memory no place. My sources are postwar Otomean humor gametes Published in Istambul, the warming capital of the Ottoman empire, while it was occupied by Allied forces immediately after world war 1. The future sovereignty of the city was unknown and there was no coherent state structure. As normative (and also subversive) popular media, human Gazettes illustrate the reverberation of postwar geopolitics with the lived and reimbursed processes of urban place. Ethno-nationalist Turkish belonging Istanbled was form of urbanization composed of place-based norms of behavior and a commonly understood cultural geography of the city. Satirical depletions of Urban Turkish and Greek encounters during the armistice era betray a Turkish anxiety surrounding territorial and historical claims to the city and also a simultaneous questioning and hardening of the imagined geographies that demarcated Turkish and Greek identities as nationally district. This research illustrates the topological and .relational dimensions of ethno nationalist identity formation and the role of urban cultural processes in political belonging in the contemporary Middle East"[18].

Geopolitical Concerns

At the heart of the geopolitical concerns today are questions about life the sustainability of life, the quality of life and the biological capacity and residence of life. In Kurdish South East Turkey, increased governmental investment in gendered development highlights the role of households, neighborhood and community production and reproduction in processes of securitization and nation building. These events suggest a deeply corporeal geopolitics at play in Turkey Kurdish question, one that rests on the intimate relationship between social reproduction and geopolitics. This article draws now specific practices and ideas around motherhood, marriage and mobility and rights in the city create and challenge ethno-national identities. In doing so, I contented that the Turkish question and understandings of Turkeishness and Kurdishness are embodied, reified and contested in the spatial constitutions of "life' s work[19].

Analysis

1. It is noted that geographers have contributed nineteen important research papers at the international level touching different aspects economic and geographic locations and situations. It is further noted that geographers should have to interact with economists as far as possible in order to identify and recognize international economic and geographic identity and further to find solutions to many posd challenges before geographers and economists.
2. It is noted that geographic research should not be conducted in isolation but it should have to be interdisciplinary one 'Economy' should be studied in relation to its geographic location and environment. 'Environmental studies' should be carried out both by economists and geographers . For example Environmental Economics and environmental geography may find good potential if we conduct the research inseparably and jointly.
3. Agrarian' studies should be conducted both by the economists and geographers besides environmental scientists in order to identify

agrarian problems, issues and farmers deaths in specific locations of the space.

4. Liner discriminate analysis may be used along with quantitative statistical methods for analyzing the research topic in-depth.

Gaps:

1. Study on Human - Mangrove interaction.
2. Role of livestock in a growing economy with spatio-temporal trends.
3. Human environmental interaction.

Tasks Ahead

1. Land Resources of sustainable development - A study
2. Land-use change - Resource base and pattern of development
3. Land-use and GIS methods for identifying economic locations & situations
4. Rural land-use change and occupational diversification
5. Spatio -temporal evaluation: population-causes for growth and migration
6. Migration and emigration: Some aspects of Economic Geography.
7. National economy and state economy: A comparative study
8. Remote sensing application for Rural Development study - Nation, State and District.
9. Environmental pollution: A study of economy in relation to socio-economic and impact on rural development.

Impact Assessment

Pollution - soil/land, air pollution/water pollution and solid waste management study and its impact on economy - A study.

Conclusion

It is identified that along with the profile of economic geographic research, there is every need to identify the urban scenario i.e urban geography international trend in geographic research. The author has made an attempt to identify the multiple problems of urban economy, urban society and urban infrastructure. Hence, the urban scenario' has been presented in the fourth chapter.

It is further identified that along with the profile of economic geographic research, there is every need to identify the urban scenario (ie) Urban Geography — International trend in geographic research. The author has presented the 'Urban Scenario' in the fourth chapter.

Notes

1. Ian Burton, "Theoretical Geography". (Quantification in Human and Economic Geography), Canadian Geographer, VII. 4, 1963, pp. 15-159.
2. Wenfu Peng Jiening Zhou, Lugunwen, Shang Xue and Lijun Dong, "Land Surface temperature and its impact factors in Western Sichuan Plateau, China", Journal of Geocart International, Vol. 32, No. 8, 2017, pp. 919-934.
3. Connor Joseph Cavanagh, "Enclosures dispassion, and the green economy: New Contours of internal displacement in Liberia and Sierra Leone", AGR Asian Geographical Review, 2017 Original articles.
4. Zhenyu Zhang, Xiaye Lin, Jim Peterson and Wendy Wright, "Cool Temperate rainforest and adjacent for ests classification using airborne LiDAR data," 'Area', Vol. 43, No. 4, (December, 2011), pp. 438448), publishing: Wiley on behalf of The Royal Geographical Society (with the Institute of British Geography). Stable UeL:http ://www.jstro.org/stabIe/41406027.
5. Kunjia Yuweil.i, Shunlim Liang, Hiangqin Wei, Xihan Mu and Yunjuw Yao, "Fractional Vegetation Cover estimation based on soil and vegetation lines in a corn-dominated area", Geocarto, International, Vol. 32, No. 5, 2017, pp. 531-540.
6. Guhia Conchedda, Eric F. Lamin and Philippe Mayanx", "Between land and Sea: Livelihoods and Environmental Changes in Mangrove Ecosystems of Senega!", pp. 1259-1284, Annals ofthe Association of American Geographers, Washington, D.C., Vol. 101, No. 6, (November, 2011).
7. Narumasa T Sutsumidoo, Paul Harris & Alexis Comber, "The Application of a Geographically Weighted Principal component Analysis for exploring twenty three years of Goal - Population change across Mongolia", pp. 1060-1074, Annals of the Association of American Geographers, Washington, DC, Vol. 107, Issue 5,2017, published online 31 May, 2017.
8. Elizabeth Havice & Limn Compiling, "Where Chain Governance and Environmental Governance Meet: Interim Strategies in the Canned Tuna Global Value Chain", pp. 292-313, published online, 21 March, 2017, U.K., Economic Geography, Vol. 93, Issue No. 3, 2017.
9. Francis Onditi & Joseplime Odera, "Gender Equality as a means to women empowerment? Consensus, Challenges and prospector Post-2015 development agenda in Africa", African Geographical Review, Vol. 36, 2017, p. 146-167 .
10. Xia Li, Guangzhao chen, Xiaooping Lin, Xun Liong, Shaojian Wang, Yamim Chen, "A New Global Land-use and Land cover change product at a I MK Resolution jar 2010 to 2 100 Based on Human Environmental Interactions", Annals of the American Association of Geographers, Washington, D.C., Vol. 107, 2017, Issue No. 5, pp. 1040-1059, 2th April, 2017.

11. Jiag Jun Qiao, Jay Lee, Xinyue ye, "Spatiotemporal Evolution of Specialized Villages and Rural Development: A Case Study of Human Province, China", Article October, 2015, Annals of the Association of American Geographers, Washington D.C.
12. Daniel Haberly & Darin sz Wojcik, "Earth Misreported: Centralization and Variegation in the Global Company Network", pp. 241-266, Published online: 22 December, 201 6. U.K. Economic Geography. Vol. 93, 2017, Issue -3.
13. Gavin Bridge & Michael Bradshaw, "Making a Global Gas Market: Territoriality and production Networks in Liquefied Natural Gas", pp. 215-240, Published online, 22 March, 2017, Economic Geography, Vol. 93, 2017, Issue No. 3.
14. Ostry A.S. et.al. Mt Joccup, "Environ Health", 1OC4), 368-374, October-December, 2004, more; Labour Markets and Employment in security; Impacts of Globalization on Service and Health care - Sector.
15. Luis F. Alvarez Leon & Colin J. Gleason, "Production, Property and the Construction of Remotely Sensed Data", Annals of the American Association of Geographers, Washington D.C, Vol. 107, 2017, Issue No. 5.
16. Andrea Ballatore, Mark Graham & Stickled Sen.", Digital Hegemonies: The Loealness of Search Engine Results", pp. 1194-1215, Annals of the American Association of Geographers, Washington D.C, Vol. 107, Issue No. 5, 20 I7.
17. Ryan E. Baxter & Kirby E. Calvert, "Estimating Available Abandoned Crop Land in the United states: Possibilities for Energy Crop Production", pp. 1162-1178, Published online 21 April, 2017, Annals of the American Association of Geographers, Washington, D.C. Vol. 107, 2017, Issue No. 5.
18. Amy Mills, "The Cultural Geopolitics of Ethnic Nationalism: Turkish Urbanism in Occupied Istanbul (19/8-/923)". pp. 1 179-1193, Published online, 21 April, 2017, Annals of the American Association of Geographers, Washington, D.C. Vol. 107, Issue No. 5, 2017.
19. Jessie Hanna Clerk, "The Life of the state: Social Reproduction and Geopolitics in Turkey's Kurdish question". Published online 17 June, 2016, Annals of the American Association of Geographers, Washington, D.C, Vol. 106, 2016, Issue No. 5.

Chapter - 4

Urban Geography: A Survey (International Geographic Research)

'Urban Geography' is the study of city or town and its geographical characteristics which have an impact upon its structure, growth, and developments and the opposite structure of rural life in every aspect in the total framework. In simple terms it relates to a civilized mechanical life tool. 'Urban Geography' is defined as the "study relates to the space, structure and functions which are dissimilar to the rural scenario besides, the essential study of Urban spatio-temporal Geography and further study of the Urban Challenges and find solutions to the present [current] problems of urbanites". (Quotation Minc Please).

The author attempts to present the current scenario of geographical research in urban geography and its elaborative content is further presented with the following aspects in all respects. They are: (1) Urban Ecology and Environment, (2) Urbanization, (3) Methodology- Globalization & Technology, (4) Urban Growth - Urban Sprawl, (5) Urban Poverty, (6) Immigration & Migration, (7) GIS Methodology, (8) Culture, (9) Gaps in Literature, (10) Suggestions for future areas of research to young geographers.

1. Urban Ecology and Urban Environment

The increasingly pervasive phenomenon of light pollution spans several different fields of concern, including loss of the night sky, energy wastage and the effects of artificial light on circadian rhythms and nocturnal ecology. Although the scale of the problem has grown significantly in recent decades, the underlying dynamics remain only partially understood beyond the identification of specific technological pathways such as the rise of Light Emitting Diodes (LEOs) or the capitalist transformation of the nocturnal realm. It is suggested that current approaches to the study of light, including the identification of "Urban Atmospheres", the elaboration of existing approaches to urban ecology or the extension of "Smart City", type discourses, do not capture the full complexity of the politics of light under late modernity[1].

Urbanized Lands

The author in this study develops an informed modeling approach that follows a bottom up planning strategy to define plausible urban growth scenarios. Landscape aesthetics suitability of the area was generated using multi criteria evaluation method in this case. Then, a buffer zone of IKM was considered to extract the average values of aesthetics suitability scores surrounding urban patches with medium physical size (10-30 hectares). The averaged values were considered as the dependent variable. In the next step, landscape matrices of these urban patches, as explanatory variables, were also computed to measure compositional and configuration based attributes of urban clusters. Bivariate associations (Pearson Correlation analysis) and statistical relationships (Linear regulation algorithm) between landscape matrices and their associated aesthetics values were measured and modeled. According to the results, both composition and configuration values are significantly correlated to the dependent variable in which configuration-based attributes depicted a stronger explanatory power[2].

Urban land cover

Reliable and up-to-date urban land-cover information is valuable in urban planning and policy development. Due to the increasing demand for reliable land-cover information, there has been a growing need for robust methods and datasets to improve the classification accuracy from remotely sensed imagery. This study assess the potential of the newly launched landset 8 sensors thermal bands and derived vegetation indices in improving land-cover classification in a complex urban landscape using the support vector machine classifier. The study compared the individual and consumed performance of landset 8's reflective, thermal bands and vegetation indices in classifying urban land-use cover. The integration of landset 8 reflective bands, derived vegetation induces and thermal bands over all produced significantly higher accuracy classification results than using traditional bands as standard one (i.e. overall user and producer accuracies). An overall accuracy above 89.33% and Kappa index of 0.86, significantly higher than the obtained with the use of the traditional reflective bands as a standard one dataset and other analysis stages.

Overall, the results of this study indicate that the integration of the landset S's OLI and TIR data presents an invaluable potential for accurate and robust land-cover classification in a complex urban landscape, especially in areas where the availability of high resolution datasets remains a challenge[3].

2. Urbanization

Gran's agglomeration has resulted in an important peripheral urbanization over the best two decades, followed by a considerable increase of distance travelled and an explosion of passenger car use as the principal transportation mode, to the detriment of public transport use and walking. A political will has emerged to promote public transport use by the implementation of a multi nodal public transport system (bus, tram and metro). Unfortunately this policy has failed to address sustainable mobility development and the promotion of soft mobility, where working, as a pillar of urban mobility is marginalized in public spaces development. The author aims to analyze the negative impact of the increase of passenger car use on the model reversal of daily practice displacements and on public spaces development. Likewise, it examines the areas earmarked for walking in the new projects of public transport promotion through the study of the first tramline project in Gran, Algeria[4].

Environmental Governance / Urban Geography

"This article concentrates on how hydro-social relations are differentially structured across technical experts engaged within diverse and multiple networks of institutional and bureaucratic practice and the implications thus has for more inclusive forms of environmental governance and decision-making empirically focus on stimulator governance in Chicago and Los Angles as a means to capture the range of geographical and institutional variations in environmental knowledge. Both cities face considerably different water resolve challenges in the United States but are at the forefront of developing comprehensive and progressive urban water governance reformist, hydro-managerial, hydro-rationalist and hydro-pragmentist. Each of these represents a particular understanding of how hydro-social relations should proceed. They all align around shared framings of integrated management and the utilization of the best available science and technology to drive decision-making. Consensus, however, masks fundamental differences among the varying groups of expertise. Differences center on the perceived effectiveness of different types of infrastructural interventions, of market and economic incentives, and the role of new institutions and rules to govern storm water. I argue that each frame looks to structure hydro-social relations to fit their own vision but consequently offer apolitical strategies that reduce water quality and quantity problems to their technical and hydrological components[5].

3. Methodology

"This paper presents a methodological approach to estimation of urban population using the volume of simple houses and high-rise residential buildings obtained from an 1 KONO S-2 Ortho-image and light detection and ranging (Lieder) data. The estimates are directly executed at the first scale level (i.e). The housing unit) and are then aggregated at the census district level for further validation with the aid of official data supplied by the local and several governments. Unlike prior works, this study executes a thorough assessment of horizontal and elevation accuracy for the 1KONO S-2 and the construction of a 3D city model, the accuracy assessment of the Ortho-image and digital surface models (PSMS), and the quantification of urban population. The validation was accomplished by means of linear regression and associated hypothesis tests, considering the estimated population and the reference data. The results showed that there was a systematic underestimation of population. An average, the conducted estimated assessed 31 fewer inhabitants per district and lie 1.350/0 below the expected values given by the reference data. In spite of the observed under estimation, the estimated population can be regarded as equivalent to the population provided by the reference data at a 1% level of significance[6].

Globalization & Technology

Against the backdrop of modernity entering its digital stage, theoretical conceptions of the relation between technology and space regard the two as becoming increasingly interlinked. In human geography, such conceptual advances are mainly reflected either in models of Globalization as being driven by technological change or in more specific investigations into technologies that have an inherent relation to space and place such as locative India. The article opens up a geographical perspective on the relation of technology and space by focusing on the micro level of the everyday use of technology. A fram e-work is proposed that understands the symbolic geographies involved in the everyday interaction with technology as a combination numeric components (Places and networks coded by numbers), deictic expressions and graphics (as in egocentric interfaces sensitive to the users location) and common to eponyms (established place names). Not only is this model applicable to the example of locative media, it also promises to be transferable to various kinds of complex infrastructure systems (including those for communication and transport) and media technology. The last part of the paper reflects on the apparent resistance of traditional geographic concepts (such as place, landscape, region or nation state) in the context of digitalized life words. This can be explained by innovations in interface design enabling both providers and users of technological systems to integrate familiar geographic concepts into interaction language[7].

4. Urban Growth

The way cities are evolving and expanding has been a challenging issue and several urban growth models who proposed. Most of these models have common objections but they differed in terms of implementation or theoretical assumptions. Urban growth modeling projects that simulate urban spatial change, across the globe, have been reviewed in this study with a view to providing reference for future urban-growth modeling development. The mathematical concepts of the modeling techniques were critically examined for holistic understanding of their key components and also to priority a broad description of the existing urban growth modeling works, which has always been ignored in the literature[8].

Refugees & Settlements

In 1967, the Government of the Republic of Zambia (GRZ) began resettling Angolan refugees into independent Zambia's first official refugee camp, Mayukwaynkwa Q.,I in Kaoma District, Western Provincc, Zambia. After nearly 40 years living as refuges, Angolans began a process of repatriation or return - to Angola. By 2010, the prospects of full repatriation from Mayukwayukwa to Angola were limited, forcing the community, the governments and the international refugee regime to address long standing appeals for alternative settlement options. This article begins by providing an historical understanding for refugee-hosting in Western province, Zambia and precedes to trite the even evolving contours of protracted refugee-hosting at Mayukwayukwa refugee settlement. The author has utilized in-depth interviews to assess in Zambian and Angolan perceptions of refugee-hosting settlement repartition, citizenship and identity. Ultimately, the protracted displacement to Angolans in Zambia challenges the discourse that privileges repatriation as the ideal durable solution. Additionally in this case, refugee settlement not only illustrates the utility of geographic approaches to refugee research, but also contributes to broader discussions regarding geographies of mobility, transnational identities, development and displacement[9].

Resettlement Issues

The impact of immigration on Canadian cities since the Second World War has been profound, especially following the removal of barriers to non-European immigrants in the 1960s and the significant increase in the number of immigrants admitted since the mid-1980s. Over the million immigrants internal Canada in the 1990s and the Vastmajority have settled in just a few metropolitan areas. As a result, the social geography of large Canadian cities has been transformed an issue that has attracted considerable attention from Canadian geographers. In this paper researcher on these changes - published by cultural, social and urban geographers between 1996 and 1999 - is surveyed. This work is exceedingly diverse in emphasis and method, and has contributed a great deal to our understanding of the relationship between immigration and urban change, particularly in the areas of housing, the labour market and neighborhood life. In general, the geographers are emphasizing the complexity of outcomes, highlighting on the one hand the importance of local contingency and on the other the growing connections between Canadian cities and global processes. This research challenges traditional theories of immigration and urban structure and in so doing will redefine the way they conceptualize urban spatial structure and urban social life[10].

Urban Sprawl

Swimming pools are together an exemplificative outcome of urban sprawl and an indicator of socio spatial polarization in metropolitan regions. A comparative analysis of the spatial distribution of pools in three Mediterranean cities (Barcelona, Rome and Athens) provides and alternative reading of recent urbanization in southern Europe, questioning the supposed homogeneity in socio-economic patterns and processes across the region. The socio-spatial structure underlying disposed urban expansion in these three cities was studied using 53 background indicators at the spatial scale of municipalities. Four indicators were proposed to study variability in the spatial distribution of pools. Relevant differences between cities were observed in the density of pools, reflecting heterogeneous patterns of dispersed urbanization and class segregation, socio-economic polarization in Athens, settlement scattering and social mix in Rome and a more balanced socio-spatial structure in Barcelona. The spatial distribution of pools in the three cities was found associated with different socio-economic factors, outlining the role of spatial inequalities disparities between high income and low income neighbourhoods, the diverging economic base and place specific attributes. Results from Barcelona, Rome and Athens case studies provide contrasting views of the relationship between class segregation and the local socio-economic context. The resulting patterns of sprawl reflect distinct urban models belying the supposed homogeneity of the 'Mediterranean City' archtype [11].

Urban Workers and Immigrants

Recent studies identify a robust positive correlation between productivity of urban workers and the presence of a diverse range of immigrants in their midst. Seeking to better understand this relationship, this article tests the hypothesis that the rewards from immigrant diversity will be higher in metropolitan areas that feature more inclusive social and economic institutions. Institutions ought to matter because they regulate transaction costs, which, in principle, determine whether, or non-diversity offers and advantages or disadvantages. We exploit longituditional linked employer-employee data for the United States to test this idea, we triangulate across two measures that differently capture the inductiveness of urban institutions findings offer support for the hypothesis. In cities with low levels of inclusive institutions, the benefits of diversity are modest and in some cases non-existent, in cities with high levels of inclusive institutions, the benefits of immigrant diversity are positive, significant and substantial. We also find that weakly inclusive institutions hurt natives' considerably more than foreign born workers. These results confirm the economic significance of immigrant diversity, while suggesting the importance of local social and economic institutions[12].

5. Urban Poverty

"In this article we argue that the recent focus on the suburbanization of poverty is problematic because of the ambiguities and consistencies in defining sub-urban. To improve transparency, explicability and comparability, we suggest the research on the geographical changes to the distribution of poverty should focus on three questions: (1) How centralized is urban poverty? (2) To what extent is it decentralizing? (3) Is it becoming spatially dispersed? With respect to all three questions, the issue of quantifying uncertainty has been under researched. The main contribution of the article is to provide a practical and robust solution to the problem of inference based on a Bayesian Multivariate conditional autoregressive (CAe) model, made accessible via the R software package CAR Byes. Our approach can be applied to spatiotemporally auto connected data and can estimate both levels of and change in global relative centralization index (RCI) local RCIs and dissimilarity indexes. We illustrate our method with an application to Scotland's four largest cities. Our results show that poverty was centralized in 2011 in Glasgow, Dundee and Aberdeen. Poverty in Edinburgh, however, was centralized: Non-poor households' trend to live closer to the center than poor ones and increasingly so. We also find evidence of statistically significant reductions in centralization of poverty in all four cities. To test whether this change is associated with poverty becoming more dispersed, we estimate changes to evenness and local decentralization of poverty, revealing complex patterns of changes"[13].

Rural - Urban Interaction

"In tropical countries, urbanization and rural deagriculturalization are intimately linked processes. Growth of the construction industry, which draws labour and materials from rural and peri-urban areas, is a key component of rural-urban interactions. Using a case study from the Perote Valley, Mexico, this article analyses how urban driven concrete aggregate mining affects rural agricultural lands and livelihoods. Land-cover changes associated with aggregate surface mining were examined by comparing aerial photographs, satellite images and Global positioning system field maps of the Valley. Besides, sixty formal household interviews were conducted in three agrarian communities. Results indicated that between 1995 and 2006, mined surface area quadrupled from 181 to 706 hectares and the spatial distribution of surface mines followed a predictable pattern of extraction, from most to least accessible. Numerous factors persuaded farmers to convert to mines, including increased prices for aggregates, consumerist desires for new homes and declining returns to agriculture. Moreover, formers believed they could or would re-cultivate mined land following extraction. Yet, in 2006, among the three agrarian communities 202 (45 per cent) of the 482 hectares of agricultural land extracted were not re-cultivated. Although deep mines were likely to be abandoned, findings suggest that land revalorization occurred among farmers, with the scarcity value of non renewable aggregates gradually surpassing the land's agricultural value. In the long-term, diminished agricultural potential resulting from wide spread mine abandonment might drive farmers to migrate to cities for employment, creating a positive feedback between urban demand for construction aggregates and mining of agricultural fields,,[14].

"Despite some improvement since the early 1990s high proportions of recently - arrived migrants (in Australia) remain dependent on unemployment benefits. Contrary to most analyst's exportations, there is evidence that disadvantaged migrants are becoming more, rather than less, residentially concentrated"[15].

6. Immigration

This paper examines the role of guaranters who help immigrants with the immigration process and with establishing new lives in Rural Africa. The Kaonde live in Mufumbwe District, North Western province, Zambia. There are four other ethnic groups in this area: the Luanda; the Luvale; the Chokwe and Luchazi. These four groups are immigrants and are permitted by the Kaonde Chief to live in the Kaonde area. There are three basic types of migration. "Village to village", where people move from one village to another "Viacity", where people from villages move to the city for work and later move to a village and "city to village", where people born in the city move to a village. Most non-Kaonde immigrants moved to village to escape family conflict or to avoid creating energy of their urban life styles among their relatives. They rely on guaranters, who are known in the Lunda language as Chinsahu. Any person can become a guarantor if he or she is trusted by an immigrant. Guarantors can help to ensure building new lives in rural Africa[16].

Urban Health

It's aimed to examine spatial clusters of urban deprivations using five dominants and the influence of the observed pattern on household health Status in Rorin, Nigeria. Data were obtained through a household survey of the 20 works of Ilorin metropolis. A systematic sampling method was used to select one out of every 10 houses in randomly selected streets in each ward while one household was selected through a simple accidental method of first contact in each building. In all, 800 households were sampled. A composite Index of Multiple Deprivation (IMD) was obtained by aggregating scores for each domain. Multiple correlation analysis was used to determine the relationship between the IMD and household health status. Deprivation in housing quality offered significant explanation (75.7%) for the variation in household health status (R=0.87) (bzO.05), public health policies in the post-2015 sustainable development Goals must consider housing characteristics within the city, classify areas according to needs and determined areas, within the city, that require positive discrimination in health resource allocation [17].

The uncertain geographic context problems, Annals of the Association of American Geographers, 102 (5), 958-968, refers to the problem that findings about the effects of area-based contextual variables on individual beheviours or out-comes may be affected by how contextual units (e.g. neighbourhoods) are geographically delineated and the context to which these areal units deviate from the geographical context. It is a significant methodological problem because it means that analytical results can differ for different delineations of context units even if everything else is the same. Drawing upon Kwan (2012) Kwan, M.P. 2012. The Uncertain geographic context problem. Annals of the Association of American Geographers 102(5): 958-968 - (Taylor & Francis Online) (Web of Science (R)) [Google Scholar] and recent social science studies (especially environmental health and neighborhood effects research), this article further elaborates on the nature of the UGCOP and explores how recent advances in geographical information system (GIS) and geospatial technologies can help address the problem. It discusses possible means for mitigating the UGCOP, especially with regard to the collection of detailed individual space-time data with global positioning system, construction of individual activity spaces, and the use of qualitative and web-based GIS to capture people's activity location and everyday experiences. It also further discusses the challenges for future research that seeks to address the UGCoP[18].

7. GIS Methodology

The present study adopts an integrative modeling methodology, which combines the strengths of the SLEUTH model and the Conservation Assessment and Prioritization System (CAPS) method. By developing a scenario-based geographic information system simulation environment for Haghtpar city, Iran, the manageability of the landscape under each urban growth scenario is analysed. Besides the above, the CAPS approach was used for biodiversity conservation suitability mapping. The SLEUTH model was implemented to generate protective urban layers of the years 2020, 2030, 2040 and 2050 for each scenario (dynamic factors for conservation suitability mapping). Accordingly, conservation suitability surface of the area is updated for each time point and under each urban development storyline. Two way analysis of Variance and Duncan's new multiple range tests were employed to compare the punctuality of three scenarios. Based on results, the managed urban growth scenario depicted better results for manageability of the landscape and less negative impact on conservation suitability values[19].

Remote Sensing & GIS in Urban Studies

"Based on containing the spatial extent of urban expansion, the urban development boundary concept provides guidance on resource constraints and urban policy development for urban areas and aims to meet the new demands of urban development under the background of a new type of urbanization in China. We applied remote sensing and geographical information system (GIS) techniques, along with slope, land/use, exclusion, urban extent, transportation and hill shade (SLEUTH) model, to identify urban growth boundaries in Changzhou City, China. We then comprehensively considered various land-use regulation policies and the conveying capacity of landresource to construct an urban development boundary model. This model was tested using empirical data on the delineation of flexible and rigid urban development boundaries. We argue that China's position as the largest developing countries in the world has resulted in significant uncertainties in its socio-economic development; therefore, the construction of Chinese Cities requires both flexible control and a rigid management structure. The model developed in this study successfully meets the construction of needs of China's Urban Development, Particularly as it contains an optimal degree of generalizability[20].

Tecnology & Economy

"Technological and structural changes in the Australian economy have led to a decline in unskilled and semi-skilled employment and this has had a market effect on labor market opportunities for immigrants. This paper reviews the labor market experience of humanitarian programme arrivals and considers the policy implications of high levels of unemployment among this group. It is suggested that humanitarian obligations do not end with entry to Australia and it is the interests of the receiving society and humanitarian programme arrivals for greater public investment in skills developed to help improve labor market outcomes among this group[21].

IT Information Technology and Software Services

Offshore has become a staple term in the lexicon of economic globalization, and yet, it has also become a bit of a black-box term. The range of meanings attached tends to mask the complex, dynamic and emergent qualities that invited for the contrast reinvention of offshore spaces. This article examines the activities of Russain transnationalizing entrepreneurs of Russian information technology (IT) offshore - a sector that is made up principally of firms based in Russain specialized in providing

IT and software services for clients located in Western Europe and North America - and analyses how their unique experiment to expand operations to Ukraine is generating a new spatial reality. One with internal hierarchies and subspaces, which I call the offshore.

Drawing on the narratives of engineers, managers and directors of Russian firms who drive the creation of this new off-offshoring reality, the article focuses on the process and practices developed by transnationalizing entrepreneurs. I document how these actors devised the initial attempts for expanding their operations in Ukraine, how they proceeded to implement and adopt their plans, and what language-from biological metaphors to pragmatic business terminology they use to grasp the newness of this process, this article highlights various unexpected difficulties that the firms encountered in their expansion efforts and discusses the unique new multidimensional knowledge work space that becomes the reality of such firms as a result of experimental rehizomatic techniques and practices that they develop - such as management circulation, multidirectional and reversed training and the development of online forums and in the fresh groups to encourage enthusiasts based in the new locations[22].

8. Culture Qatar Pride / Urban Geography/ Global visibility

"Qatar has been projecting power through services of the spectacles, investments and interventions. These include the new Doha Skytime, ownership of the tallest building in Europe (London 's shared); Al Jazeera media, involvement in the Libyan and Syrian civil wars; and in postnatal, hosting global events such as the 2022 Soccer world cup, Qatar also holds 14 per cent of all known natural gas reserves and boasts the World's highest per capita income. Our article relates to Qatar's global visibility and presence to processes of power and accumulation in its capital city, Doha. We do this through a focus on migrant workers who, in this highly urbanized state, make up 89 per cent of population. Their lives and labours provide a window on the relationships between different stages of power and accumulation. The spectacle and the work and labour that sustain it. Interesting geographies of foreign labour and the urban space tackle require secreting through sharpened critical and post colonial lenses on diversity and urban modernity[23].

9. Gaps in Literature

Important gaps in research are identified at the international level and there is every need to the researcher to plan the multidimensional development in the existing urban environment particularly "Urban infrastructural facilities". There is every need to provide the infrastructural facilities to the urban population, keeping in view of the growth structure of urban population. Spatial variations in urban structure leads to a low quality of life. Quality of life is much depend upon quantity of the inputs of facilities.

Step to be taken up by Government

1. Complete urban survey is to be taken up and it should be computerized.
2. Urban infrastructure should have to be provided in the small and medium towns i.e., "Rurban". The selected towns should be provided basic infrastructural facilities.
3. The urban survey relates to the following aspects. Such as:
4. How many people are residing in a house/settlement
5. What is the income level i.e. sources of income.
6. Lavatory facilities are available or not?
7. What is the status of the street? Road facility CC Road and drainage system, street lighting including the availability of the community Hall and other urban infrastructure.
8. Providing the employment scheme such as - self employment etc to achieve this, there is every need to provide training for self sustainability in a given urban environment which leads to integrated city/urban development over a period of time.

10. Future areas of research

1. Metropolitan planning & development - A City above 1 crore population.
2. Smart City Development: A Case Study
3. Spatial patterns of urban development in India/State
4. Classification of urban land-use/urban land-space
5. Woes of urban voter's water problem: Analysis of Urban voters mechanism.
6. Urban population Estimation and planning social/urban over heads or integrated, urban development.
7. Geospatial based urban growth models and sub-modeling.
8. Migration, immigration & Emigration process in a metropolitan city - A geospatial study and strategy.

9. Socio-economic disparities and urban sprawl - A case study in South India/North India [A comparative analysis of socio-economic and urban history].
10. Inclusive Urban Growth - Growth Modeling: A Geographical Study
11. Urban Poverty Rural Poverty - Exodus of Rural - Urban Rurban spatial structure: Socio-economic and geographic study
12. Changing occupational structure of Urban population - A Historic and Social Study of a city
13. Land-cover changes using Satellite images, Aerial Photography, Remote Sensing & GIS Study.
14. Sustainable Urban Health: Health Care delivery Systems.

Conclusion

It is concluded that the researcher has to find an innovative topic for the research study and untouched areas should be touched for current research. In chapter-five Rural Geography, Regional Planning and Development Planning aspects are presented for fresh look for implementation.

Notes

1. Matthew Gandy, *Negative Luminescence",* pp. 1090-1107, published online 11 May, 2017, Annals of the American Association of Geographers, Washington D.C, Vol. 107, 2017, Issue No. 5.
2. Yousef Sakieh, Abdolrassoul Salmaumahiny, Syed Hawed Mirkarimi & Sepidah Saeidi, *"Measuring the* relationships between landscape aesthetics suitability and spatial patterns of urbanized land'), An informed *modeling frame work for developing urban growth scenarios",* Vol. 32 N. 8 pp. 853-873, 2017, Journal of Geocarto international.
3. Terence Darlinton Mushore, Onisimo Mutauga, Jolino Odindi and Timoths Dube, "Assessing the Potential of Integrated Land sat 8 thermal banks" with the traditi onal reflective bands and derived vegetation indices in classifying urban land scopes". Vol. 32, No. 8, pp. 886-899, Journal of Geotlarto, International, 2017.
4. Asmaa Kerrouche & Mohamed Madani, "The Marginalization of Walking, the Achievished of Sustainable Mobility Policies in Gran (Algria)", pp. 236-252, African Geographical Review, vol. 36, 2017, Issue No. 2.
5. Joshva J. Cousins, "Structuring Hydro-Social Relations in Urban Water Governance", pp. 1144-1161, published online, 12 April, 2017, Annals of the American Association of Geographers, Washington, D.C, Vol. 107, 2017, Issue No. 5.
6. Liva Tomas, Leila Fonseca, Claudia Almedia, Fernaudo Leonardi & Madalma, "Urban Population Estimation based on residential buildings, volume using

IKONO S-2 Images and Lieder data", pp. 1-28, published online 14 December, 2015, International Journal of Remote Sensing, Volume 37, 2016, Issue Supl.

7. Tilo Felgenhauer, "There-imagining of the geographical tradition in the digital age", pp. 1-12, Geographic Annler, Series B, Auman Geography, 2017.
8. Sukuman Ibralion Musa, Mazlan Hashian & Mold Nadzvi Md. Reba, "A Review of Geospatial-based Urban growth models and modeling initiatives", Vol. 32, No. 8, 2017, pp. 813-833, Journal of Geocarto International.
9. Angela G. Subulwa, "Settlement, Protracted displacement, and repatriation at Mayukwayukwa in Western Zambia", African Geographical Review, Vol. 32, 2013, Issue one, pp. 29-:43.
10. Daniel Hiebert, *"Immigration and Changing Canadian City",* The Canadian Geographer, Vol. 44, Issue No. 1, March, 2000.
11. Ilaria Zarnbon, Pere Serra, David Sarre, Margherit Carlucci and Luca Salvati, *"Byond the Mediterranean City: Socio-economic Disparities and Urban Sprawl in three Southern European Cities"*, pp. 1-19, Geographic Annular, B. Human Geography, 2017.
12. Tom Kemeny & Abigail Cooke, *"Urban Immigrant Diversity and Inclusive Institutions",* pp. 267-291, published online: 30 March, 2017, U.K. Economic Geography, Vol. 93, 2017, Issue No. 3.
13. Leokavanagh, Duncan Lee & Gwilym Pryce, *"Is Poverty Decentralizing? Quantifying Uncertainty in the Decentralization of Urban Poverty",* Annals of the American Association of Geographers Washington, D.C., Vol. 106, 2016, Issue No.5, published online: 16 September, 2016.
14. Malthew Fry, *"From Crops to Concrete: Urbanization, Re-agriculturalization. and Construction Material Mining in Central Me)(icCo"*. Annals or the Association of American Geographers, Vol. 101. No. 6, (November, 2011), pp. 1285-1306.
15. EXCERPT. Publ.Me.12320563 Welfare Benefits and Residential Concentrations, Amongst Recently, Arrived Migrant Communities.
16. HARA Masaya, *"Migration Process for Immigrant in Rural Africa: Focusing on Guarantors in a Multieconomic community in North Western Zambia"*. E-Journal Geo, Vol. 12, 2017, No. 1.
17. Usman A. Raheem, *Understanding the Spatial Context of Sustainable Urban Health in Africa for the SDGS: Some lessons from the condors of deprivation in Rorim. Nigeria"*. pp. 216-235, *African Geographical Review.* Vol. 36. 2017. *Issue No. 2.*
18. Mei-Po Kwan, *"How GIS can help address the Uncertain geographic context problem in social science Research"*, Annals, GIS, Vol. 18, 2012, Issue 4, pp. 245-255 USA.
19. Mehdi Sheikh Goodarzi, Yousof Sakieh & Shabnam Mavardi, *"Measuring the effect of an on going* urbanization process on biodiversity conservation suitability index: integrating scenario-based urban *growth modeling with conservation assessment and prioritization System (CAPS)"*, Vol. 32, 2017, No. 8, pp. 834-852, Journal of Geo-Carto, International.
20. Penghuni Jiang, Qianwencheng, Yuan Gong, Liyan Wang, Yunqian Zhang, Liang Cheng, *"Using Urban Development Boundaries 10 constrain Uncontrolled Urban*

Sprawl in China", Published online: 02 August, 2016, Annals of the American Association of Geographers, Washington, D.C., Vol. 106, 2016, Issue No. 6.

21 Stevens, c., Asia Pac Migr J 6(2), 185-212.1997.more Balancing Obligations and Self Interest: Humanitarian progamme settlers in the Australian Labor Market.

22 Malanie Feakins, *"Coff-Coffshoring from Russia to Ukraine, How Russian Transnational Entrepreneurs Created a Post - Soviet IT Offshore"*, pp. 397-417, Journal of Economic Geography, Vol. 93, Issue-4, 2017, U.K.

23 Robina Mohammed & James D. Sidaway, *"Shards and stages: Migrant Lives, Power, and Space viewed from Doha-Qatar"*. Published online, 31st August, 2016, pp. 1397-1417, Annals of the American Association of Geographers, Washington, D.C., Vol. 106, 2016, Issue-6.

Chapter - 5

Rural Geography, Regional Planning and Development Planning (International Level)

Geographical factors play an important and predominant role in the origin, growth and development of a 'region'/ A 'region' is that part of the earth's surface with natural character besides its acquired attributes such soil, agricultural, industrial, administrative, urban and so on. 'Regional planning' is quite necessary for the development in every aspect such as agriculture, industrial, urban and administrative development over time and space. Development planning or planning for development is an innovative scientific programmes adopting various methods of technology for its upliftment from its natural built-up and the author has presented various aspects of rural geography, regional planning and development planning research studies with the following frame-work and format. They are: (1) Cultural elements, (2) Population, (3) Methodology, (3) Migration, (4) Millennium Development, (5) Cluster Development and Regional Innovation, (7) Gaps in research, (8) Future area studies.

1. Cultural Elements of Geography

"This article uses data on two vernacular traces, to progress and business names to explore and analyze cultural geographies of the U.S. Appalachian region. First, cluster detection is applied to a set of generic place names (toponums) that are found to be components of a relatively unique Appalachia geographic vernacular. The results from this exercise are mapped into common spatial representations of the Appalachian region, where they are combined with a selected marker of perceived regional identity - business that have the character string "Appalachia" in their name to construe a study area for subsequent analysis. Finally, data are collected for places inside of the derived study area to examine determinants of Appalachian identity in zero-inflated count regression model. The findings convey important insights about specific heritage and environmental variables that correlate with and lively contribute to the production, patterns of Appalachian identity in the United States. In addition, the conceptual model and methodology of the article are claimed to be transferable to other studies of cultural requires, heritage and identity, both inside and outside of Appalachia 2016 Annals of the Association of American Geographers Journal R.G. Impact & Description-Research Gate - Impact Rankings (2015, 2016 and 2017).[1]

Regional Geography

For many students, entering a geography degree programme exposes them to new, probably unexpected, and material. Almost all beginning undergraduates in United Kingdom University, for example, have studied geography right through high schert but although some subject matter and approaches are common at both types of institutions much is not, university degree programmes introduce theories, philosophies and methodologies, especially in human geography, not even implicitly - encountered in their previous studies. By contrast, many who enroll in American geography graduate programmes have not studied the subject at either high school or during their undergraduate careers. Each group therefore needs foundational exposure to contemporary geography's breadth and depth of approaches[2].

2. Population Estimates

"Social Scientists regularly relay on population estimates when studying changes in small areas over time. Census treat data in the United States are a prime example, as there are substantial shifts in threat boundaries from decade to decade. This study compares alternative estimates of the 2000 population living within 210 tract boundaries to the Census Bureau's own retabulation. All methods of estimation are subject to error; this is the first study to directly quantify the error in alternative interpolation methods for U.S. census tracts. A simple areal weighting method closely approximates the estimates provided by one standard source (the neighbourhood change data base), with some improvement provided by considering only area not covered by water more information is used by the longitudinal Tract Data

Base (LTDB), which relies on a combination of areal and population interpolation as well as ancillary data about water covered areas. Another set of estimates provided by the National Historical Geographic Information Systems (NGHIS) uses data about land-cover in 2001 and the cement road network and distribution of population and housing units at the block level. Areal weighting alone results in a large error in a substantial share of tracts that were divided in complex wages. The LTDB and NHGIS perform much better in all situations but are subject to some error when boundaries of both tracts and their component blocks are redrawn. Users of harmonized tract-data should be watchful for potential problems in either of these data sources[3].

Population and Mobilities

"In the 1990's the population (of New Zealand) is experiencing higher levels of mobility than at any time in its history. However, with regard to European migration to New Zealand, the settler flows of the past have been overtaken in importance by reverse flowed of temporary migrants. Now flows of new settlers come predominantly from Asia and Pacific. This paper is about return migration - a process that has been largely ignored in the literature on international migration to New Zealand[4].

Land Rights

Recent work in indigenous geographies theorizes that indigenous peoples are differently placed in relation to the state, neoliberal development and to other indigenous peoples, rather than a simplistic indigenous /non-indigenous people and custodial/exploitative frame. Indigenous people's claims to land Australia are overlain by a reductionist, settler legal topography that confers land rights to save aboriginal people, native title to others and nothing to those whose connections to land have been 'washed away in the tide of victory'. This paper deals with finding from interviews and focus groups with Aboriginal Australians in settled Australia, following negotiations over resource development. The authors argued that in the often-distant periphery where land rights and native title are compelling, the notion of intercultural space is exotic, marketed and celebrated. In contrast in the clearly demarcated settler zone, intercultural spaces with significant Aboriginal populations are steadfastly inscribed as dispossessed space, ripe for development. Because of neoliberal state - corporate alliances and interests and the ease with which the development agenda is progressed in settled Australia, they argued that realization of where, now and why these geographies have come about is needed to achieve national intercultural space. Without this understanding, only a limited intercultural space will be achieved[5].

Health & Sanitation

Slums are examples of localized communities within third world urban systems representing a range of vulnerabilities and adoptive capacities. This study examines vulnerability in relation flooding, environmental degradation, social status, demographics and wealth in the slums of Accra, Ghana, by utilizing a place based approach informed by fieldwork, remote sensing, census data, and geographically weighted regression (GWR). The study objectives are threefold: (1) to move slums from a detachment into a continuous classification and examine the spatial patterns of the gradient, (2) to develop measures of vulnerability for a developing World City and model the relationship between slums and vulnerabil ity and, (3) to assess if the most vulnerable individuals live in the worst slums. Previously developed slum index is utilized and four new measures of vulnerability are developed thorough principal components analysis (PCA) including a novel component of health vulnerability based on child mortality visualizations of the vulnerability measures assess spatial patterns of vulnerability in Accra. Ordinary least squares (OLS), spatial and GWR model the ability of the slum index predict the four vulnerability measures. The slum index performs well for three of the four vulnerability measures, but is least able to predict health vulnerability, underscoring the complex relationship between slums and child mortality in Accra. Finally, quintile analysis demonstrates the elevated prevalence of high vulnerability in places with high slum index scores[6].

The spa town of Kihosaki, within Toyooka City, Hyogo prefecture, is one of the main hot-spring areas in Western Japan. At the center of Kihosaki are the public baths, where most small - scale notices are located. These attract young adult and female visitors from neighbouring prefectures. Besides this, the numbel of tourists from both Asia and Western countries are increasing. The recent boom is a result of tourism-based community design undertaken from the late 2005. This study clarifies the characteristics of the tourism businesses, hotels and visitors in Kinosavi in recent years, focusing on the interpersonal relations of leaders who promote community design combined with tourism. The results indicate that male leaders in their 20's to 40's promote community design combined with tourism in Kinosavi. It is notable that residents who do not have occupations related to the tourism industry are included among the leaders. Historically creating indoor spas within hotels resulted in conflicts within the community. However, a mutually beneficial spirit emerged under the tourism based community design theme. Traditional festivals contribute to establishing personal relationships among various ages and occupation and male leaders in their 20's to 40's were entrusted with festival organization. Recently, women have joined the community design combined with tourism efforts, and neighbouring areas have cooperated with Kihosaki[7].

3. Methodology

The paper clarifies elements of regional attractiveness on subjective evaluations using data on 1,000 municipalities in "Regional Brand Research 20 IS". First, 75 variables of regional characteristics in the research are summarized into 13 elements of components analysis. Second elements of regional attractiveness are clarified by multiple regression analysis in the subjective attractiveness as an explained variable and the 13 elements are explanatory variables. As a result, it was clarified that subjective attractiveness is composed of 11 elements. The element of tourism and leisure is the most effective in subjective attractiveness, following food and agriculture, convenience of living and shopping and history. The elements clarified in the study will make it possible to evaluate defective regional attractiveness[8].

Civil Conflict

This article examines how fictional representations of civil contlict and its consequences enable contemporary African writers to capture the raft of way in which war remains a troubling/agent within the national body. This is especially relevant in the context of post colonial societies living the traumatic aftermath of decolonizing and liberation wars. Focusing in the figure of the 'child soldier' as an emblem of the experience of displacement as engendered by civil war Chimamanda Ngozi Adichie's Half a yellow sun (2006) and Chris Abanis Song of Night (2007), I argue that the figure of a child Soldier qua displaced individual brings into relief a polemic and politically explosive commentary on the state of nationalism, statehood and citizenship in post colonial, post independence Africa[9].

4. Migration and Migrants

Drawing on the thresholds approach, a model that in corporate geography and mobility studies to understand migration from the perspective of migrants, this article examines the importance of location and route(s) in determining the journey of Somali Migrants. We use the protracted displacement of Somali to address the process through which actual and perceived geographies are produced and reconfigured by Somali migrants on the move. The doing so we look at how different migration partly connected through transnational net works, influence the geographical perception of migrants at the core of any cross-border experience. Employing ethnographic field work with sources in South Africa and the U.S., we argue that the personal subjective realities of migrants are instrumental in determining when, how, and where to migrate, even under conditions of forced migration. Besides flows of economic and social remittances, narratives shared through networks have a central role in shaping views of time and space in migration. Locations previously perceived as ideal find destinations sometimes become temporaries, while sites of transit are remembered as sites of belonging and investment. This comparative piece demonstrates the importance of these multifaceted flows as new geographies of displacement and migration emerge, renewing and challenging notions of place and transnational horne[10].

Immigration

"This paper examines the impact of immigration on the relatively depressed labor market in South Australia during the 1976-81 inter causal period. The findings indicate that recently arrived immigrants have experience exceedingly high unemployment rates. However, this is consistent with the view that immigration generates additional job opportunities for Australian residents (except in the manufacturing industry and trades occupations), and that immigrants experience high unemployment rates specifically because they have been unable to compete a 'fair share' of jobs away from residents already entrenched in the labour market"[11].

Refugees and Rehabilitation process

Israel has since 2005 become an important destination for asylum seekers from sub-saharan Africa. Even though it was an early signatory to the convention relating to the status of Refugees the development of an Israeli asylum regime has only gathered pace since then. That regime is underpinned by a discourse of securitization that regards most asylum seekers as hostile ' infiltators' . This paper is based on fieldwork among the Eriterian refugees community in Isreal and in using two concrete case studies analyses the struggles of Eriteran refugees to realize rights and lay claim to a viable future. Their struggles provide insights into the wider debate on the back of concentrate footing of universal rights in actual political space[12].

Social Welfare payments

"There has been much concern about the level of social welfare payments repeatedly received by New Zealanders in Australia. Analysis of these payments showed that the recent renegotiations of the Australian and New Zealand social security agreement will shift much of the financial burden to the New Zealand Government. Recent migration patterns show increasing shares of non New Zealand-born persons and older persons amongst the New Zealand Citizens arriving in Australia[13].

Millennium Development Goals

The millennium Development Goal for water sought to halve the number of people without-access to an improved source of water. The World officially achieved this goal in 2010, but a significant body of literature critiques the definitions and measurement access. This paper provides an overview of this criticism and uses data from Dares-Salaam to show that poor quality, fluid price structure, unreliable supplies and burdensome collection all limit water access even though they are not considered in official measurement. This paper further argues for a meaningful definition of access is required to guarantee real development and poverty reduction[14].

Millennium Development as a process of Development

Sanitation has been a core-development - related keyword since the Millennium Development Goals were launched, but its improvement in sub-Saharan Africa has been considered generally slow. So far, sanitation needs have been mainly addressed technically and economically while lacking proper intersection with related conditions, such as health education, cultural and environmental contexts, gender and ownership. These elements seem how to be considered by the new sustainable goals launched in 2015. More emphasis is given to the importance of providing differentiated, instead of homogenized, guidance to any process of change and material intervention, including sanitation projects. These cannot be reduced in terms of external environmental engineered cycle connecting households but have to be valued for the way they involve people's bodies, ecosystems and livelihoods. This paper presents the results of a research conducted in Kenya and Particularly in Taita hills an area mainly served by pit latrines and hit by environmental degradation. The research was meant to understand local perceptions and attitudes toward implementation of different types of ecological sanitation solutions that make possible the establishment of a closed loop of nutrients connecting food production and sanitation. The findings indicate the importance of local cultures and Personal Preferences in defining sanitation choices particularly in rural areas, starting by the consideration of local livelihoods and preexisting systems serving the human waste disposal cycles[15].

6. Cluster Development Regional Innovation

This paper deals with the evolution of the leT cluster in southern Sweden. It forges a link between the literatures on cluster development and regional innovation systems (RISS) and explores both conceptually and empirically how cluster evolution is shaped by regional context conditions, including changes in the policy support structure and the knowledge base configuration of the RSI. The authors used qualitative research methods and descriptive statistics to analyze the development and transformation of the ICT cluster. CIuster emergence was enabled by strong analytical and synthetic knowledge bases in the region. The global crisis and changing geography of the ICT industry triggered a transformation of the cluster towards a new growth path with stronger symbolic components. Cluster transformation was facilitated by policy actions that promoted a diversification of knowledge bases in the region. The strategy was to combine existing competences in ICT with new competences in media and design and to develop new activities around the theme of new Media, which integrates analytical, synthetic and symbolic knowledge. They showed that the endowment of the RIS of a variety of knowledge bases and their combination has essentially supported long-term cluster development and transformation[16].

Environmental Justice Problem / Social advantages & disadvantages

"We examine the role of the regulatory state in the inequitable distribution of social advantages and disadvantages. To illustrate this, we examine the spatial distribution of exposures to air taxies from noxious land users (commonly referred to as the environmental justice problem) and inquire into the nature of state action that would allow such inequality. Findings from an inquiry lead us to focus more closely on the administrative functions of the state, especially its role as a regulation body. A case study focusing on health risks from incompatible land uses illustrate how spatial inequalities result from the formally neutral rule-making actions of regulatory agencies and their particular organizational cultures. We describe the ethical basis of the regulating state in terms of its formal juridical, deontological under pinning. In contrast to this stands the alternative ethical concept of care, which is inherently relational, contextual and preferentially, attentive to the needs of the vulnerable. We argue that the regulatory state can be reformed; building structures of case to better address issues of spatial inequality. We end with a discussion of how the institutional model of the caring state might be achieved in practice'[17].

"The broad objective of this paper is to assess the impact of crop husbandry production on desertification in Kenya's and Semi-Arid environments. The classification of existing crop area by type, including tree and agricultural crops.

1. The impacts of intensive Crop husbandry and existing faming systems on desertification.
2. The identification of the potential use of specific crops for desertification control in relation to the possible agro-forestry systems and
3. The paper proposes some appropriate monitoring strategies required for the improvement and sustainable crop production systems and environmental management in Kenya.
4. The strong impact of man on the fragile arid and semi-arid ecosystems in Kenya significantly contribute to bad lands and causes of desertification.

The most recent formation of bad lands started with the sedenterization of the nomadic pastoralists during the colonial administration.

A plan to intensify the development of African agriculture and a national crops policy probably enhanced desertification process in the ASAL.

However, the impact of such plans and policies on desertification have not been closely monitored with a view to improving environmental management in crop production systems in the ASAL.

The author opined that, "We need efficient methods of monitoring and assessing desertification in the ASAL. But the question is who can carry out the monitoring mechanism in the ASAL efficiently[18].

Natural Hazards

"The process of desertification which is a result of different environmental and human efforts includes several essential part processes.

Not all these processes are indispensable for desertification but most of them play smaller or bigger roles whenever and wherever desertification takes place.

The author has dealt with the following aspects for description and specification.

They are:

1. Salinity and secondary salinization in desert agriculture.
2. The main aspects of secondary salinization and Alkalization.
3. Annual Salt Accumulation in the catchment of two rivers.
4. Processes other than irrigation causing secondary salization
5. Salt regime of salt Balance.
6. Depth and chemical composition of ground water
7. The techniques of irrigation & calculation of salt balance.
8. Salinity hazard; and
9. Scheme methods recommended for the control of salinity and Alkalinity in Irrigated Areas.

Finally, opined that, "one of the main tasks for the future is intensify the study of the possible hazard of increasing irrigation on soil and the environment and to elaborate up to date and economical methods for the combat against the growing danger of the development of harmful processes'[19].

Economic Globalization

"The objective of this investigation was to determine the impacts of economic globalization on labor markets and outline potential pathways for these changes to affect health status in industrialized nations. A systematic review of the economic globalization and health literature revealed that, under the impact of globalization and market deregulation, the past 25 years have witnessed de-industrialization, shifts to nontraditional, in secure work agreement and relatively high levels of unemployment most developed nations. This has occurred in the context of hyper mobility of capital, relative immobility of labor and declining market position for unskilled labor. Such structural changes in the labor markets in conjunction with shifts in educational opportunities and requirements have resulted in the increasing marginalization of unskilled workers from the labor market. Aside from direct effects on health due to the threat and experience of unemployment, and given that income inequality within nations is a main driver of national health status, lowered relative wages for the unskilled will probably affect national health status through increased income inequality's"[20].

7. Gaps in Research

1. Vernacular traces and cultural geography - A strategy of development
2. Geographical thought.
3. Population Mobility
4. Health and Sanitation
5. Tourism

Future areas of Research

1. Recognition and identification of vernacular traces - A study in cultural geography.
2. Contributions and Development of Geography: A strategy for geographic ideas and development thought.
3. Rural population migration - Migration Patterns and occupational mobility - A Geographic study
4. Rural health and Health care delivery systems: A case study of South India/North India - A comparative study
5. Tourism - A Geographic Study for integrated tourist places and circular system in Tourism.
6. Migration - Time and space study.

Conclusion

1. It is concluded that there are identified gaps in literature (research) and these gaps should be filled in with current research activity.
2. Occupational mobility --A Geo-environmental study of migration for occupation- search for livelihood and their patterns.

Finally, it is concluded that Bibliometric Analysis of International geographic research - an evolution and evaluation and current need is to know the trends in geographical research. Keeping in view of it, in chapter-6, the above analysis has been presented in order to know the areas of research genetically and further attempts mey be made by the researches.

Notes

1. Russel Weaver, Chris Holtkamp, *"Determinants of Appalachian Identity, Using Vernacular Traces to study cultural geographies of an American Region"*, Annals of the Association of American Geographers Journal R.G. Impact & Description Research Gate-impact Rankings, (2015, 2016 and 2017).
2. Ron Johnston, *"A Students Introduction to Geographical Thought: Theories, Philosophies, Methodologies"*. Geographical Review, American Geographical Society, New York, 11 November, 2016.
3. Joson R. Logan, Brain J. Stults & ZengWangXu, *"Validating Population Estimates for Harmonized Census Tract Data 2000-2010"*, Published online; 17 June 2016, Annals of the American Association of American Geographers, Washington, D.C., 2016, Issues No. 5.
4. J.M. Lidgard, N.Z..J. Geogr (97), 3-13.4, 1994. more.
5. Talia Bohl-Van den Boogaard, Jennifer Carter & David Hollinsworth, *"Calise we 're gonna lose bigtime': Indigenous resource negotiations in settled Australia". pp. 1-15. Geographic. Annaler. B.Human Geography.* 2017.
6. Marta M. Jankowska, John R. Weeks & Ryan Engstrom, Dothe Most Vulnerable People live in the Worst Slums? A Spatial analysis of Accra, Ghana, Annals of GIS, Vol. 17, 2011, Issue 4, pp. 221-235.
7. IWAMA Kinuya, *"Development of Tourism - Based Community Design in the Spa Town of Kinoshi in Toyooka, Hyooka, Hyogoprefecture, Focusing on Interpersonal Relations, E-Journal CEO,* Vol. 12 *(2017)* No. 1 p. 59-73, *of Group Leaders.*
8. TANAKA KOICHL *"Analysis of Elements of Regional Attractiveness Based on Regional Brand Research"*, E-Journal GEO Vol. 12 (2017) No. 1, pp. 30-39.
9. Tonysimes dzsilva, *"From Lived Spaces to literacy Spaces: The Figure of the Child Soldier in* Contemporary African Literature", African Geographical Review, Original Articles, 2017, pp. 1-11.
10. Mamie Shaffer, Ginlia Ferrato & Zaheera Jinnah, *"Routes. Locations and Social Imaginary: a comparative study of the ongoing production of geographies in Somali Forced Migration"*. African Geographical Review. 2017. original articles.
11. Harrison, D.S., Econ Rae 60 (168), 57-67.3 1984 more The Impact of Immigration on a Depressed Labor market: The South Australian Experience.
12. Tanja R.Muther, *"Realising rights within the Israeli Asylum regime: a case study among Eritreali refugees in Tel Aviv"*, African Geographical Review, 2017, Original, pp. 1-12.
13. Rapson, V., People Place 4(2), 34-45, 1996 more.
14. Sarah L. Smiley, *Defining and Measuring Water Access: Lessons from Tanzania for moving forward in the post-millennium Development Goal Era"*. African Geographical Review, Vol. 36, 2017, Issue 2, pp. 168-182.
15. Matias Auderson & Paola Minoia, *"Ecological Sanitation: a Sustainable goal with local choices. A case study from Tait Hills. Kenya"*. pp. 183-199, African Geographical Review, Vol. 36, 2007, Issue 2.

16. Roman Martin & Michaela Tripple, *"The Evolution of the CT Cluster in Southern Sweden - Regional innovation Systems. Knowledge bases and policy actions"*. Geographic Annaler, Series B, Human Geography, 2017, Germany, pp. 1-16.
17. Raul P.Lejauo & Richard Funderburg, *Geographies of Risk, the Regulatory State, and the Ethic of Care"*. published online: 24 May, 2016. Annals of the American Association of Geographers, Washington, D.C., Vol. 106, 1016, Issue No. 5. (pp. 1097-1113.
18. Cleaphas Lado, *"Impact of Crop Husbandry Production on Desertification in Kenya"*. pp. 109-164, (in) perspectives in Resource Management in Developing Countries, Volume-3, Ecological Degradation of Land Edited by Baleshwar Thakur Concepts international Series in Geography-5, published by Concept Publishing Company, New Delhi, 2008.
19. Szablocks, *"Solemnization of Soil and Water as an Important Factor in Desertification",* pp. 80-95, in perspectives in Resource Management in Development Countries, Volume-3, Ecological Degradation of Land, Edited by Balashwar Thakur, Concepts International Series in Geography-5, Published by Concept Publication Company, New Delhi, 2008.
20. A.S. Ostry. Int. I.Health Scrv 39(I), 45-57. 009 more. "Globalization and the Marginalization of Unskilled Labor: Potential Impacts on health in Developed Nations".

Chapter - 6

Bibliometric Analysis Of The Research Papers Published @ International Level: An Evaluation (Based On 5 Years Period)

'Accreditation' and 'Assessment' are two present methods through which international geographical research (IGR) over a period of time is possible, plausible and vulnerable to evaluate its progress. The research progress could be identified at one level (ie) International level. An attempt has been made extensively with an intensive effort in this chapter to evaluate the evolution process through the identified methodology and also further identify the above process, the data base has been created and classified according to its structural components, its role for which the author has conducted a survey for a period (Time) from 2013-2017 both years inclusive and exclusively included the abroad journals such as (1) Annuals of the Association of American Geographers (AAAG), Washington D.C. USA, (2) Canadian Geographer, (3) African Geographer, (4) Scottish Journal of Geography, (5) Geographic Annaler (A. Physical and B. Economic) and (6) Australian Geographer, stretching from Canada, North America, South America, Africa, Sweden, China, Asia (China & Japan) and Australia, to some extent representing the contents from the respective continents (space and time-S.T analysis) relating to Bibliometric studies have been evaluated (It is rather an evolutionary process and assessment method from the above journals from two to five (both are inclusive) Chapters. This forms as a first model, based on the created data base of case studies 2 to 5 chapter material have been described and also presented the degree of collaboration of the authorship ratio.

The second model has been created, presented based on *7th* and 8th chapters and Secondary layer study structured for the same period with two types of material separately. (i) 33rd international geographical conference material, 21-25, August, 2016, Beijing, China, Asia. and the referred Journals in particular are the following:

1. The Geographer, Aligarh (Online published material)
2. Transactions, Pune (Journal Publications), India
3. Annals of the National Association of Geographers, India, New Delhi, (Annals); and
4. Indian Journal of Regional Science, Kolkata.

Besides the conference material relating to the body of literature, and the main objective being, 'Aggregation' and 'Disaggregation' of knowledge that has been sprinkled over the 'space' and 'time' (S&T) at different special contexts and also disaggregated and disseminated the sprinkled knowledge has been presented to the students, teachers, researchers and also for the seasoned geographers to equip and accommodate themselves with the Current International Trends (CIT) in geographical research. Bibliometric study has been carried out of 85 published lengthy and theme oriented articles (referred herein as database and where they were thematically analysed and presented with bibliometric methodology (applied). At the end, the potential areas of research besides the Gaps in research have been identified and also present and suggest model topics to be chosen and taken up by young researchers for their study in each chapter at the end.

Methodology

Time changes could generally be evaluated by the library and information scientists often used Bibliometric method of research. They make use of quantitative methods for carrying-out the analysis in order to describe the patterns of publication within a specified field or a body of literature, researchers invariably use Bibliometric methods of evolution to determine the influence of single author to describe the relationship between the two or more authors or works.

'Bibliometric's' is one way called as qualitative science besides it gives an important field/subject of information science. Bibliometrics has practical value (application) in the evoluation and evaluation of historic operations and survey through quantitative statistical techniques to carry-out the quantitative analysis. The study has been described and presented in the following para-graphs in detail by the author in order to identify and recognize the space(s) and time (T) -spatio-temporal (S&T Methodology) aspects through bibliometric analysis method and thus enable to recognize the elements of international geographical research (1OR). This is essentially a space-time study and analysis.

Analysis

"Resources are sources of energy and energy gives strength to activate the activity in deed and in need" (Quotation Mine Please). It means that sources are important for human livelihood and data collection and analysis are quite important to reach the penultimate stage of a research endevour. In this process, data collection is important particularly the method of data collection and the collected data should have to be analyzed in order to identity and interpret the structural trend of the current international geographical research (this is the major objective of this pronounced and announced project).

Methodology of collection of data

(1) Major branches of Geography of traditional nature are as follows:

1. Physical Geography
2. Economic Geography
3. Urban Geography
4. Rural Geography and Regional Planning

Stage-1: At this stage, a total of 85 research papers (table no. 6.1) have been selected and studied in depth for review purpose from the journals of international reputed nature. (All are international journals and peered papers).

Stage-2: At this stage, 23 papers are reviewed in physical geography, 19 papers are reviewed in Economic Geography, 23 papers are reviewed from urban geography and finally, 20 papers are reviewed from urban geography and finally, 20 papers are reviewed from rural and regional geography.

Table 6.1: Data Analysis relating to Eighty Five Papers (International Level)

SI. No.	*Nams of the branch*	*Single Authored*	*Two authored*	*Three authored*	*Four authored*	*Five authored and more*	*Total number of papers*
1.	Physical Geography	1,4,10,12,14, 16,17,20	8,11,18,22	2,3,5,6, 13,15,21	9,9A	7,19	
		8	4	7	2	2	23
2.	Economic Geography	1,3,18,19	8,9,11,12,13, 14,15,17	3,7,16	4	2,5,10	
		4	8	3	I	3	19
3.	Urban Geography	1,5,7,9, I0, 14,15,16,17, 18,21,22	2,12,19,23	8,13	2,3	6,11,20	
		12	4	2	2	3	23
4.	Rural & Regional Geography	2,4,7,8,9,11, 12,13,14,19,20	1,15,16,17,18	3,5,6,10	-	-	-
		11	5	4	-	-	020
5.	Total no.of papers (Cumulative)	8+4+12+11	4+8+4+5	7+3+2+4	2+1+2	2+3+3+0	85
		35 41.17	21 24.70	16 18.82	5 5.88	8 9.43	85 100.00

Stage-3: As per Bibliomatric methodology the classification of international research papers are made and presented in table 6.2. The authorship pattern of the research papers are as follows and gain analysis is also shown against them.

Table 6.2: In Physical Geography five types of authorships are visible

			Gain
1.	The single authored research papers	8	8
2.	Two authored research papers	4	4
3.	Three authored research papers	7	14
4.	Four authored research papers	2	06
5.	Five authored research papers	2	08
	Total no.of papers	23	40 (40-8=32)

Table 6.3: In Economic Geography, there are five types of authorship share also visible. They are as follows

1.	The single authored research papers	4	4
2.	Two authored research papers	8	8
3.	Three authored research papers	3	6
4.	Four authored research papers	I	3
5.	Five authored research papers	3	12
	Total no.of papers	19	33 (33-4=-29)

Table 6.4: In urban Geography, there are five types of authorship' visible. They are as follows

1.	The single authored research papers	12	12
2.	Two authored research papers	4	4
3.	Three authored research papers	2	4
4.	Four authored research papers	2	6
5.	Five authored research papers	3	12
	Total no.of papers	23	38 (38-12=26)

Table 6.5: In Rural and Regional Geography, there are three types of Authorship is noted

1.	The single authored research papers	11	11	Gain
2.	Two authored research papers	5	Total no. of papers	05
3.	Three authored research papers	4	08	(24-11)=13
4.	Four authored research papers	0	24	
5.	Five authored research papers	0		

Table 6.6: Stage-4: Paper's Typology of each category are presented as follows

		Gain	
1.	The single authored of all branches put together (8+4+12+11)	35	35
2.	Double authored of all branches, put together (4+8+4+5)	21	21

(Contd...)

3.	All three authored of all branches put together (7+3+2+4)	16	32
4.	All four authored of all branches put together (2+1+2)	05	15
5.	All Five authored of all branches put together (2+3+3)	08	32
	Total no.of papers of all categories	85	135
	= (135 - 85 Total No.of papers = 50 @ Loss 62.1 percent		

Table 6.7: Ranking of Branches According to Weightages Assigned

1.	*Physical Geography*	*Single*	*Two*	*Three*	*Four*	*Five*
	Weightage	*Authored* (10) 35×10 350	*Authored* (9) 21×9 189	*Authored* (8) 16×8 128	*multiple* (7) 16×8 128	*Authored* (6) 8×8 64
2.	Economic Geography	4×10 40	8×9 72	3×8 24	1×7 7	3×6 18
3.	Urban Geography	12×10 120	4×9 36	2×8 16	2×7 14	3×6 18
4.	Rural & Regional Geography	5×9 110	4×8 45	0 32	0	
	Total	620 (47.04)	342 (25.94)	200 (15.17)	56 (4.27)	100-1318 (7.58)

Table 6.8: Cumulative totals to cumulative Index

Sl. No.	Major Branches of Geography	Author Category					Total
		1	2	3	4	5	
1.	Physical Geography	(45 .69) 350 (56.45)	(24.67) 189 (55.26)	(16.71) 128 (64.00)	(4.56) 35 (62.50)	(8.35) 64 (64.00)	776 (58.11)
2.	Economic Geography	(24.84) 40 (6.45)	(44.72) 72 (21.05)	(14.90) 24 (12.00)	(4.34) 7 (12.50)	(11.80) 18 (18.00)	161 (12.21)
3.	Urban Geography	12 (19.35)	36 (10.52)	16 (8.00)	(6.86) 14 (25.00)	(9.82) 18 (18.00)	204 (15.47)
4.	Rural Regional Geography	110 (17.74)	45 (13.15)	32 (16.00)	0	0	(14.18) 187 1318
	Total	620 (47.07)	342 (25.94)	200 (15.17)	56 (4.24)	100 (.7.58)	100.00

Gross Observations

I. Vertical Scenario

1. It can be seen from the above table, single authored papers occupy to a tune 350 (56.45) out of 620 papers weighted contributions to physical geography is the top per cent.
2. It is further seen from the table that urban geography and rural geography play the second and third place in the category of single authored papers. Economic Geography takes the last place in the major branches of surveyed one.
3. It is also observed that two authored category of papers physical Geography occupies the first place i.e., 189/342 (55.26) papers double authored papers in economic geography takes the second place but urban geography takes the last place in this category.
4. It is seen from the table, three authored papers, physical geography occupies 128/200 (64.000/0) and Regional Geography occupies the last place.
5. Four authored papers category, physical geography has been predominantly dominates the entire scenario.

II. Gross Observations - Horizontal Scenario

1. Both vertical and horizontal totals are 1318 but the category wise differs significantly. Further observations are presented in the following paragraphs.
2. Physical geography gained 350 points/766 (45.69) single authored papers and double authored papers occupy 189/776 and three authored papers are to a tune of 128/766.
3. In Economic Geography, two authored papers tops the list with 72/161 and second place goes to single authored papers.
4. In Urban Geography branch, Single authored papers occupy the first place to a tune of 120/204 and the least category is to tune of 14/204.
5. In rural and regional geography, out of 187, 110/187 occupies by Single authored, where as second and three authored papers to a tune of 45/187 and 32/183. There are no four authored and multiple authored papers visible in this category.

III. Gross Observations - Total (Horizontal)

1. It is observed that out of 1318 points, single authored papers of all categories put together, 620/138 (47.04) followed by two authored papers with a total of342/1318 (25.94).
2. It is further observed that three authored papers takes a total of 200/1318 (15.17) followed by five authored papers. The last category is four authored papers.

Degree of Collaboration

Based on table-6.1, the further analysis is as follows:

The Degree of collaboration © among the authors was derived using

Subramanian's (1983) formula which states that the degree of collaboration is the ratio between number of multiplied authored articles plus the numbers of single authored articles (Sharma, 2009). The formula is stated as follows:

C= NM

NM+N.S.

Where

C = Degree of Collaboration

NM = Number of Multiple Authored Articles

N.S. = Number of Single Authored Articles

The International Journal Articles are committed to Scholarly Excellence and dedicated to the advancement knowledge and information technology. It aims to provide geographic knowledge (original) for sharing information about research,knowledge, experiences, perceptions and wisdom with people involved in all facets of geographic knowledge and geographic information technology.

1. In Physical Geography, the degree of collaboration is

C= NM

NM+NS

$$=\frac{15}{15+8} \quad \frac{15}{23} \quad =0.6521$$

2. In economic geography

$$= \frac{15}{15+19} = 15 \qquad = 0.7894$$

3. In Urban Geography $= \frac{11}{11+12} = \frac{11}{23} = 0.4782$

In the branch of rural and regional geography

$$= \frac{9}{11+9} = \frac{9}{20} = 0.45$$

Total picture for the total number of papers degree of collaboration is as follow:

$$= \frac{35}{35+50} = \frac{35}{85} = 0.4117$$

Conclusion

H.G.: All the branches of geography are equally developed:

1. The research contributions differ from branch to branch within the selected sample of 85.
2. The multiple authored papers are much associated with technological and innovative one with RS & GIS methodology in particular physical geography, urban geography and economic geography. But, in regional and rural geography, many single authored papers with statistical and cartgraphic methods are visible.

It is concluded that all the branches of geography could not gain the equal placing and found that inequality exists taxonomically and genetically. Inter and intra variations in the paper contributions are very vividly and variedly apparent-A stable Shed Facts

Finally, it is observed that technological and innovative papers gain more importance for publication in the are international Geographical research multiple authorship is much more prevalent which is also rampant. There is every need to present the India's contribution to Indian Geographic research and it is presented in Chapter-7, for a thorough understanding of the contributions in a systematic way.

Reference:

(1) Subrahmaniam, K, (1983), Bibiliometric Studies of Research Collaboration: A Review, Journal of Information Science, Vol. 6, pp. 35-57.

(2) Sharma, Rakesh Mani (2009), Research Publication Trend Among Scientists of Central Potato Research Institute: A Bibliometric Study, Annals of Library & Information Studies, Vol. 56, pp. 29-34.

Chapter - 7

Indian's Contribution To Indian Geographical Research - An Overview

(1). Introduction

Indian Geographer's have been conducting intensive research studies over an extensive geographic area over a period of time, upon Indian landscape. Qualitative research contribution is the main contribution since 1950 by Indian Geographers. Physiographic regions, Natural regions, agricultural regions, industrial regions, economic regions, resource regions, administrative regions, urban regions, regional settlement systems and cultural regions have been spatially and functionally spread over and the status have also been in-tune to suit with the needs and aspirations of the people living in these regions. Besides this, mountains, plateaus, deltas and valleyes are the important physiographic and tectonic evidences that are visually visible with greater indivisibility. Under physical geography, contributions have been appearing in the subfields such as Geomorphology, climatology and oceanography and the economic geography deals with agricultural activity, industrial activity and the activation of resources by human efforts to affect economic life of people residing in those regions and that it has terrific, terrible and untold recorded effort upon people and their culture.

The research papers form the traditional branches of geography are as follow:

(1) Environment and resource, (2) Physical geography, (3) Economic Geography, (4) urban Geography and (5) Rural Regional Geography and Rural Development. Chapters from Second to Fifth, core Chapters of First

series are presented relating to International Geographical Research exclusively from 1-85 numbers, where as in the second series themes are presented relating to Indian's Contribution to Indian Geographical research wide Chapters 6 & the exclusively. This Chapter is an admixture of Indians participation in the 33rd International Geographic Congress, Beijing, China, 21-25, August, 2016 besides, the material has been presented in the Indian Geography journals particularly dealt with the following Journals. They are: (I) The Geographer, Aligarh, (2) Transactions, Pune and (3) Annals of the National Association of Geographers, India, New Delhi, and so on.

The entire text of the review deals with the following aspects. They are related to state level, district level, taluk level and also village study.

1. Environment and Development
2. Geographical landscape of Karnataka
3. Geomorphology
4. Micro level- Village study - Geospatial Technologies (District Study)
5. Remote Sensing and GIS
6. Urban Geography Impact of Human Intervention on Green Environment Urban Geography (District Study) (B)
7. Urbanization
8. Urban Slums. - M.P. State Studies
9. Comparative Study of Slums
10. Rural - Urban fringe
11. Land-use and Agriculture (District Study) (B)
12. Kashmir Valley
13. Crop-Diversification (State study)
14. Agricultural Development - (Taluk Study)
15. Enterprise Development (Garden Culture - Tea Cultivation North - East Region of India.
16. Sustainable Agriculture - Haryana (State Level Study)
17. Agriculture in Drought Prone Region (Hamirpur District, Himachal Pradesh)
18. Rural Geography - Rural Development Infrastructural Facilities - Health Care System - An Evaluative study; and
19. Rural and Regional Development - (Bandipore District Study)

Many quantitative techniques have been adopted in many research papers besides Cartographic Thematic mapping of variables. Correlation, regression and factor Analysis are quite important statistical methods in addition to remote sensing and Geographic information system of the current geographic research have enriched with.

2. Indian Geographical Research and the application of Biblomatric methodology

Methodology

1. All the papers are categorized into three. They are: single authored papers, 2) Double Authored papers and 3. Triple Authored papers have been presented in table-6. I.
2. 33 international Geographic Conference papers were also categorized and marked as (b) and they were presented in Beijing, China, Asia.
3. At the third stage, all the papers are given points based upon its publication pattern.
4. Analysis has been carried out, keeping in view the methodological pattern as assigned points.

All the papers were finally given assessment based on point scores such as A Grade, B Grade and C Grade. The analysis has been presented at the end of review of articles.

3. The review of the entire text has been presented in the following paragraphs, theme-wise form as environment & development

"The understanding about the concepts concerning with water security, food, energy, Climate, environmental studies, healthcare, disaster management and sustainable development with the methods of both the national and social sciences is core to the geographical research and its application to attain human security. The developmental activities that took place in the name of socio-econornic transformation of human society without taking prior care of its implications to the immediate environment have resulted into depletion of natural resources in the environment. The environmental implications have resulted into occurrences of frequent natural hazards in the form of landslides, recurrent drought and floods, cyclones, earth-quakes and many more the effects of climate change and threat to biodiversity are two major problems in the current scenario from environmental point of view. The authors pleaded that works covering various environmental issues across India which needs wider concentration in the achievement of sustainable development". They have dealt with

Natural Hazards and Disaster, Urbanization and Urban Environment, Environment and Health, Land Resources and its Utilization, Water Resources and management, Forest Resources and Biodiversity, Developmental Activity and their Environmental Implications, and Climate Change and Environment.

They concluded that, "a balanced use of resources by keeping in mind the regional disparities along with the regional aspirations in allocating the resources and recognition of geographical disparity, socio-economic dynamics of regions, diversification of economy improvement in the quality of education and promoting health and well being in cities (like smart cities) are source of key areas where attention from a planning point of views are required for sustainable future of regional development in India.'

2. Geographical Landscape - Karnataka

"The landscape of Karnataka seems to have been suitable for human habitation from the pre-historic through the proto-historic and historic period as evidence for all the three phases one available almost all over the state. In this paper, the geographical landscape of Karnataka has been discussed. The drainage of Karnataka is based on three notable rivers - Krishna on the North, the Kaveri on the South and the two rivers (North Penner and South Pennar) and the Paler on the east. The drainage of Karnataka is mentioned as settlement from the Neolithic period onwards are located close to the rivers and rivers played a crucial role in sustaining human habitation. Similarly, Karnataka enjoys the benefit of number of mountain ranges. Mountains are a source of economically viable rocks and minerals. There is evidence for the exploitation of metals and minerals since prehistoric period. Other important geographical aspects which may have had a bearing on the historical development such as soil types, flora and fauna, climate etc. have been examined".

Methodology: Archaeological data is the primary evidence for the history of Kanataka till the early Centuries.A detailed review of archeological settlements is undertaken specially with respect to the location resource utilization etc. the primary sources used to closely study the geographical setting are the District Gazetteers which provide a very clear and detailed account. The author also undertook a field survey to have firsthand account of settlements and the immediate surroundings. Finally opined that, Karnataka is well provided with natural uniform with some areas being within forest and found wealth (Maidan) while others being more suitable for agriculture (Maidans). The abundance and unbalanced distribution of these resources have influenced the course of historical development[2].

3. Geomorphology

Fluvial geomorphology, the morphometric control on hydrological response of watershed scale in the Upper Indus Basin has been presented by Momeos et.al (2012). Kopal Kumar and Dhorde (2013) estimated the run off from impervious surfaces using remote sensing data. The effect of land-use cover changes at sub-catchment levels on down stream flood peaks was analysed by Sanyal et.al (2014) and Vashist and Bam (2013) desired a function for Raminchawn spring to predict the spring water availability during the recession period. They have dealt with the sub-fields of Coastal Geomorphology, Glacial Geomorphology, Tectonic Geomorphology, Weathering and Soils, and Geomorphic Hazards. The Atlas on "Land-forms of India from Topomaps and Images" (Vaidyanadhan and Subba Rao, 2014) provides 60 examples of different types of land forms occurring in different sections of India and resulting from different geomorphic processes such as fluvial, Aeolian, glacial, marine, tectonic weathering, mass movement etc.

The author opined that fluvial Geomorphology and Coastal Geomorphology continue to dominate the Indian Scene. The interest in tectonic Geomorphology and quaternary Geomorphology is growing and there is a scope for initiating systematic studies in other less studied sub-branches of geomorphology, such as Aeolian geography, desert Geomorphology, Karst Geomorphology, periglacial Geomorphology and mountain Geomorphology that is also an urgent need for greater attention from Indian geoscientists on India's geological and Geomorphologic heritage on geo-heritage[3].

District Study

4. Micro Level - Village Study

Detailed analysis of SONDH village in Mewat district (Haryana) by fulfilling various objectives like collecting and compiling household level data for preparation of comprehensive data base for village utilizing geospatial approach for building structure and land mapping improving their socio-economic conditions by exploring possibility of using semi-automatic tools/machines to increase productivity of existing products and deriving new products at nominal prices followed by few pictures showing interaction meeting with villagers.

Geospatial Approach

1. Building structure mapping using high resolution remote sensing satellite image.
2. Building-use and land-use mapping through ground verification by fields surveyors.
3. Attribute table/fusion table generation for identification of occupation/profession of SCs and OBC of Sondh Village".

5. Remote Sensing and GIS

"In the present time, series of different remote sensing satellite from space agencies across the World (different countries along with India) made rapid growth in both the sensors and platforms which has provided motivation for remote sensing and Geographical information system (RS & GIS) in different aspects of environmental science and other areas. The high resolution of remote sensors has evolved to capture the five details of the earth's surface features in remote sensing data. There are a no.of environmental variables, soil and surface variables, associated with mineral activities that are to some extent detectable easily with satellite earth observation data. Remote sensing and GIS can play an important role in assessment of the activities occurring on earth surface; enhance the understanding about the environmental problems and management for the better future.

Nath (2014) has presented the basic and advances of remote sensing and GIS in a Book entitled "An Introduction of remote sensing". He has covered different aspects of the use of remote sensing processes in disaster response, land-cover, geological structure and dynamic earth system.

Urban and Regional planning and history of urban growth and urbanization reveals that urban areas belong to the most dynamic land-cover types of the earth, urbanization is a process, which is manifested in the growth of population living in towns and cities. This" process is inter-linked with the economic advancement and involves the multiplication of point's of population concentration as well as increased in the size of individual urban concentration. The urban growth towards fringe areas has considerable impact on the environment; natural resources, land-use and land-cover and vegetation cover at large. It is necessary to impact analyze land-use and land-cover of urbanization and its surrounding and offer ways and solve the emergent problems while at the same time create option for future development.

Shekhar and Diddee (2015) have captured transformation of Balawadi Village of Pune from rural to urban through the topographical maps and the satellite images for visual understanding and found that the whole process of urbanization in the village through 4 steps. This study tried to explain the process that lead to spatial and temporal changes that happen in an urban fringe.

Further presented land/land-cover, land resource degradation and management, it gives information about the natural and socio-economic factors, human livelihood and development and also this pattern of any region is an outcome of various Physical Cultural factors and their utilization by the human in time and space, Disaster Monitoring and Mitigation Environmental changes water resources/water quality/water recharge, vegetation cover/forest management, mountain region/glacier, slope etc could be possible through remote sensing/GIS[5].

District Level Studies through Remote Sensing and GIS Methodology

6. Remote Sensing

Increase in population and associated activities of urban and industrial development have resulted in exerting more pressure on natural resources, thereby causing damage to the ecosystem upon which human sustenance rests. It has resulted in degradation that cannot be set right by the self regulations mechanism of the environment. The present study through satellite based analysis involves various aspects such as land-use, water resources, land degradation, etc to map and monitor the changes due to anthropogenic activity in a part of Sonbhadra district, South of river Son proximity of coal and abundant water resources has given impetus to the industrial development and resultant urbanization of the region. Thermal Power Generation, Open Cast Coal Mining, Limestone quarrying and stone crushing for building materials are the important activities which is causing adverse impacts on the assignment of the study area. An in-depth environmental analysis of the study are points to the drastic changes in general land-use of the region on account of undue anthropogenic interventions thereby lowering the overall quality of the geo-environment has been highlighted in the present study. This study emphasizes the importance of remote sensing based studies in environmental analysis of an area of interest[6].

Urban Geography

"Urban Geography has emerged as one of the major sub-fields of Geography studies Spaninra-City setting as well as inner city scenarios. The studies, however, are majorly at a descriptive best an analytical level, very few studies endeavor to predictor forecast for the future urban problems have been the focus of many of the studies. These relate to lack of availability of water to issues related to transportation, fuel used for cooking in the houses and the impact on women's health. Urban islands as they are related to land-use, density of built up areas, population size, urban geometry, density of built-up area, anthropogenic factors besides others.

She has dealt the aspects such as (1) inclusive city concept, (2) Micro climates, (3) Urban Islands, and (4) Air quality particularly land-use - Ramesh (2011) in 'Land-use in Suburban: Perunagalethur village of Chennai Metropolitan area' has made a temporal study of one of the villages on the fringe of Chennai, which has evolved into a suburbs. Various economic, social, political and technological factors have facilitated this development over a period of time and it manifests in one land-use pattern of the settlement. The author has identified changes in the land-use over[5] phases of time and the predominant finding is that of conversation of agricultural and follow-land to resolution and institutional line. Further presented the aspects such as (6) Migration and slums, (7) Urban Transportations, (8) Urban Hinterland, (9) Water ways, (10) Waste Disposal, (11) Urban Environment (12) let in Urban studies and challenges of urbanization[6].

She concludes that studies pertaining to the urban hinterland relate to topics as diverse as of artisans in the NCT of Delhi, spatial modeling of urban growth and urban influence, evolution of hinterland besides others and studies have also viewed the water crises in various cities, management of solid waste, impact of domestic waste disposal on health and innovative use of GIS in City Sanitation planning. A number of studies pertain to demography and some have made use of ICT in studying Urban Problems[7].

7. Urbanization

"Ashok. K. Dutt & et.als. using a large number of variables and correlating them with each other conclude that, "Socio-Economic correlates of Asian Urbanization" suggest that increasing urbanization is linked to higher economic development that also leads to a decline in fertility and infant mortality, it has to be pointed about that rural poverty. That always existed - is not enough of a trigger to intimate and perpetually feed the rural - urban migration stream. The expanding economic base, especially the service sector with the latter's growing share in the national economy, has been an important contributor to urban growth (as seen in references fig. 7.2). An important insight one gets from the paper the author 's think of increasing urbanization as problem, and retention of the rural population in the country side by encouraging village industries, a solution. This is highly debatable, with excessive and disproportionate overload of population in rural areas in India, where agriculture accounts for 14-150/0 of the GOP, migration to high-income areas is not a regressive phenomenon. Settlement systems in general and urban centres in particular are driven by contemporary development and economic realism[8].

M.P. State, Capital

8. Urban Slums

"Improving quality of life in the slums is one of the development Goals for new millennium. The existence of slums is essentially a manifestation of poverty, along with the economic and industrial development. Slums will continue to exist and grow further increasing number of slums constitute a major challenge to development. In India, about 6.54 crore population live in slums, which are in appalling condition due to dilapidated and infirm housing structures, poor ventilation, overcrowding, inadequate lighting, lack of safe drinking water, absence of toilet facilities, unavailability of basic services. Supply of these services is one of the biggest challenges to all urban planners and policy makers. Bhopal, being capital of Madhya Pradesh has huge influx of population who resides in slums and squatter settlements. According to Census 2011, the total area of Bhopal city is (287.17 Sq.Km) with (17.96 lack) population . There are 380 slums distributed in Bhopal City with 491 lakh (27.38%) slum dwellers living in miserable condition slums are spread all over the city, core and middle (old Bhopal) area city has huge no.of slums though periphery has less than in comparison (New Bhopal). This study is based on primary survey conducted in ten selected slums distributed in different zones in the city. For this purpose 200 hundred households were surveyed. There is a need to conduct a deep analysis of the socioeconomic condition of slums in Bhopal city and also examine the problems and challenges related to infrasrructural facilities across the slums, it will also helpful to come out with more practical policy implication at the grass-root level[9].

Two Cities

9. Comparative Study of Slums

"This paper aims to highlight the different and similarities between the slums of two cities i.e., Allahabad and Varanasi, slums of Allahabad and Varanasi resembles certain common traits like zonal location of slum demographic characteristics, occupational structure of slums, area occupied by slum, tenure, dwelling density, availability of basic facilities etc and also differs on some aspects like morphology of slums notification status, ownership of land, area, surrounding slum etc. this study reveals existing gap in policy making and identifying the specific needs of slums in both cities[10].

10. Rural - Urban Fringe

"Low literacy and disparity in the literacy between male - female, between rural urban and between young and aged population are one of the most important issues in the developing countries of the World. India, as a developing country is not only facing the problem of low literacy but also facing the wide disparity between male-female and between rural - urban literacy. This paper analyze the spatial variation in the literacy rate, male - female literacy differential and changes in male - female literacy as well as in total literate population of rural - urban fringe of Faizabad city. This study is based on secondary sources of data obtaining form the primary census abstract (2001, 2011) Census of India. The study reveals that highest growth of male as well as female literacy and low male-female literacy differential is found in primary fringe in comparison to secondary and rural fringe literacy rate is continuously increasing from 2001 to 2011 in rural - urban fringe, but literacy rate is most significantly increasing in primary fringe than secondary and rural fringe. Female literacy rate has much improved than the male population from 2001 to 2011 in rural urban fringe. High literacy, less male-female literacy differential in primary fringe is because of various reasons:- (1) Influence of the city, (2) Nearness of the city, (3) Improving the educational facilities, (4) Increasing accessibility, (5) Increasing awareness. The study concluded that there is an inverse relationship between distance from the city and literacy rate[11].

Study of Taluk in a District Framework

11. Land-Use and Agriculture

"Land is the most important among the natural resources of the country.

Land-use is mainly seen for urban and rural activities. Main components of the rural land-use is the Agricultural land-use, land-utilization survey provides relevant information related to agriculture that is the crops and crop land-use patterns. The basic aim of the land-use study is to know and record the distribution of land under various uses in different socioeconomic and environmental conditions.

The author has dealt the aspects such as land-use, land-cover changes cropping pattern, land-suitability and agricultural production, Agriculture Typology and Agriculture Development.

1. Kumar and Ranjan (2014) have studied 'Land-use modelling changes in Godavari river basin - A comparison of two districts in Andhra Pradesh. This study was conducted between 1985 and 2005. The Application of Agents modelling reveals that the accuracy in land-use change in major classes like agriculture, forest and urban areas is higher than in the other classes.
2. Geethanjali (2015) made a temporal study of change in cropping pattern considering 2011 data, selecting nine crops namely, Cholan, Paddy, Pulses, Fruits & Vegetables, Cotton, Groundnut, Coconut, Gingerly and Sugarcane to analyze the spatial changes in cropping pattern and crop combination in Dindigul district of Tamil Nadu state.
3. Land suitability & Agriculture Production, Kumar, Masilamiani and Nagara Thinam (2013) studied land-suitability following the land-suitability index, land capacity by considering Rainfall. slope, soil texture and applied to study land-use/land-cover changes and their variations.
4. Agricultural Typology: Chendrayudu, Mamatha Keeranachari and Reddy (2015) studied the Agric. Typology for food security in Rayalaseema Region: They identified agricultural types, at agricultural holding level and find out differences from one holding to another, finds the week links in agricultural among the farmers. suggested ways and means to minimize disparities and develop agricultural and achieve food security through the advantages of agricultural typology[12].

12. Kashmir Valley

"Kandi refers to an area which is upland sub-maintenance having scarcely of water,undulating topography steep and irregular slopes, and low water retentive soils with terrain dissected by numerous gullies. Production of crops is low in these belts, cultivation of maize, pulses, paddy on terraced farmers, oil Seeds, fodder plus rearing and keeping of cattle are the dominant economic activities in the Kandi areas. The type of agricultural practices chosen by an individual in Kandi belts depends not only on the geographical location, but is a function of the culture, education, technology and income of the people. The poverty exists in all economies of the World in general, however, being more prevalent in such area. The present study conducted in the rain-fed 'Kandi' areas of Kashmir Valley (Jammu Kashmir), was a primary survey based on structural questionnaire & aimed to find out the impact of small landholdings on the economy of people living in these areas. It was found that people were mostly involved with primary sector having marginal landholdings (with average landholding size or 7.27 Kanals) due to which their income was low and hence poor living standard[13].

State Study

13. Crop Diversification

"The objective of the paper is to analyse the crop diversification and cropping pattern in west Bengal in the last 30 years. It is pursed through increasing crop diversification index we have used Bobbs and Martin Indices (1960) to measure diversification in the state. The study is entirely based on secondary sources of data. The result shows that in all most all districts crop diversification is taking place and more so in the second phase of our study i.e, 2000-01. The study also reveals that the area under cereals and pulses decreased where as area under oilseeds and cash crops has increased. The hilly and backward districts recorded more diversification but towards lower value crops, in accordance with the changing water availability, weather condition, access to credit and risk. It goes against the normal feature of crop diversification where farmers diversify in favour of high value crops. The diversification has taken place in favour of high value crops or the crops which seem to provide higher relative return to the cultivators[14].

Land Use and Agriculture

14. Agricultural Development, Taluk Study

Nithya and Rose (2015) study agricultural drought vulnerability using remote sensing images, numerous indices were developed to describe the state of the land surface vegetation, with the potential to detect and monitor amenities such as drought. The result of the indices is compared with the land-use / land-cover data derived from remotely sensed images acquired for selected years under study (2005, 2009). The techniques like Normalized Difference Vegetation Index (NDVI), Normalized Difference Water Index (OWl), Normalized Difference Moisture Index (NOMI) are used to assess the agriculture drought vulnerability on Triruchuli Taluk in Virudhanagar district of Tamil Nadu[15].

15. Enterprise Development / Tea-Culture

"There is a shifting paradigm within the Tea industry in India from colonial corporate ownership to small scale individual ownership production system. It has expanded from the few traditional tea growing states to fifteen states in India in the last few decades . Almost all The states in North-East India grow tea today. As per Tea Board of India there are 1,61 ,328 members of small tea growers (STGs) with the cultivation of 1,64,396 hectares contributing roughly around 30-35 percent of tea production. In most of the cases, they are the first generation unemployed Youth into tea cultivation and enterprise development. On the other hand, the organized tea sector has 1686 tea estates with tea area of 4, 16,027 hectares. The labour relations are well defined in organized sector are also covered under the plantation Labour Act. But the unorganized sector the labour relations are evolving and new set of organizations like Boughat leaf Factories (BLEs). Tea producing Societies (SHGs), confederation of Indian Small Tea Growers Association (ClS'TA) and state level associations etc are being emerging for the safeguard of livelihoods across the country and North-East India. There are around 5.64 lakh families and 2.51 million people are associated in the unorganized sector of tea industry in India besides 1.25 million workers employed directly in organized sector of tea industry and 2.50 million people are directly dependent.

The quality of employment in organized tea industry differs from small scale cultivation which may be better termed as unorganized sector. Livelihoods in this sector seem to be more vulnerable and their engagement with the market has series of middlemen and buffer zones. The Tea Produced Societies (Self-Help Groups) are yet to be organized as viable units and establish their credentials and visibility in the market particularly in auctions. The question of un-recognition and low level of entrepreneurship skills always push them into a situation where buyer-driven market fixes the prices of their product (intermediary product i.e. green leaf). It highly reflects on the quality of their engagement, increase livelihoods and compromising with overall quality of life and unexplored potentials within the tea industry'[16].

State Level Study - Haryana

16. Sustainable Agriculture

"The author has attempted to analyze the problem of declining water table, factors for its depletion and suitable mitigation measures to combat the declining water table problem for sustainable agriculture development in the state. The major mitigation strategies include crop diversification, delay in paddy transplantation, laser leveling, zero tillage, technology, bio-drainage and rainwater harvesting for artificial ground water recharge in the aquifers".

Water resources particulars (SCM) / year), Rainfall / water table,; Haryana: Groundwater Resources and their utilization (2009), Reasons for declining trend of water has been identified, change ill cropping pattern, increase in Area under irrigation, increase in number of tube wells, Declining trend of rainfall, and declining in canal irrigated area, ground water management options crop diversification, irrigation efficiencies of various irrigation techniques (percentage), yield gain and water saving of important crops under drip irrigation, potential crop areas which can be brought under micro irrigation, comparative Rice Yield and irrigation water saving under laser and traditional and leveling adoption of Raised Bed planting technology, artificial recharging of aquifers promotion of roof top rainwater harvesting, ownership of Ground water legislation and pricing of ground water are presented.

Finally, the author opined that, to make the schemes result oriented, strong political, administrative and technical will is needed. Furthermore, there is urgent need to create awareness among the users so that ground water development and management can be a people's programme in Haryana.[17]

District Study

17. Agriculture in Drought Prone Region - Hamirpur District

The present paper analyses the level and severity of household food insecurity in drought prone region of Uttar Pradesh Bundel Khand. The study is based on empirical data collected from field survey of fourteen villages in Hamirpur district of Uttar Pradesh by using will structured questionnaire in the year 2011. The findings of the study reveals that approximately 67.38 percent of households in the study area are food insecure. The paper further investigates the severity of the food insecurity based on twenty indicators under four dimensions of food security. The paper also gives the reasons behind the high severity of food in security in the region[18].

18. Infrastructural Facilities

Health Care System

The Indian System has been greatly influenced by the changes in political and economic thinking in India as well as the World. These changes get reflected in both the changes in approach to healthcare delivery and the structure of the delivery system. The approach to health care delivery in India was top down, though great emphasis was placed on universal access and equity. The Alma-Ata Declaration in 1978 shifted focus on democratization and its coronary, the primary health care approach. This ushered in changes at the structural level as well. In the above context, the present paper deals with the changes in approach to health care provision in India since its independence and to unearth the processes and events that have led to changes in the provision and structure of health care delivery systern[19].

19. Rural and Regional Development - Koppal District

Bandipur District Study

Rural development denotes economic betterment of people and greater social transformation. Regional development refers to a quality of life in a region in terms of economic growth, social development, political maturity and environmental conservation regional development is inclusive of rural development. Rural is segment of a region. Separately it will invite injustice with urban segment of the region.

The author has dealt the following aspects in this paper such as regional disparity, regional development and rural development, and finally conclusions are presented.

Misra (2014) in his presidential address in 35th Indian Geography Congress held at Madras University highlighted that there is a very high degree of disparity in regional development. A number or regions in India have traditionally remained backward. Geographers must pay attention to those regions in depth, with a view to suggesting policies and strategies for natural resource development, human development, and infrastructure planning ideally speaking, each micro region of India should be studied in a systematic way, making micro regions building blocks for meso regionaization and further pointed out that the time has come for promoting regional studies and published regional monographs which can be utilized for further development.

Socio-economic disparities in Koppal district, a set of 32 indicators was taken and concluded that is necessary to reduce the gap of disparity of development from one taluka to another in order to reduce the disparities in the levels of agricultural development (TaUalli and Nagaraj. 2015).

Regional Development

Evaluating the concentration of growth centre for balanced regional development in Belgaum division, it was concluded that spatial distribution of growth centres in the region is uneven. It varies from one physiographic division to another. The analysis suggested that the distribution pattern of higher over growth centers should be brought towards more or less uniform distribution by upgrading another 6027 growth centres (Nayak, 2013).

Akhter (2014) in his study of 100 households in four randomly selected villages in Bandipur district of Jammu & Kashmir highlights the food insecurity in rural areas. It was found that agriculture does not produce enough foot the population of the villages studied. It resulted in food shortage and malnutrition . It is suggested that the agriculture situation in the region can be improved and food security achieved through regular supply of food from outside. The non-agricultural activities need to be promoted in areas where land is unable to support people[20].

3. Gaps in Literature (Observed)

1. Physical geography has been enhanced the status and structure of Indian geographic research and the mounting-up research papers are under the sub-fields of Geomorphology and Climatology. But, it is quite disheartening to note that Oceanographic research has been stand still. (It is a major gap in physical geographic research and there is no substitute to it).
2. Research in Geographic Thought did not find any proper place in the entire gamut of framework of current research studies.

3. Traditional branches of geography should be clubbed with RS and GIS, and it is quite satisfactory such as Physical Geography, Economic Geography, Urban Geography and Rural Geography and Rural Development.
4. Urban Poverty studies are at the minimum level.
5. Surveying methods mapping and photographic evidence is not at all satisfactory.

4. Tasks Ahead

1. Land/form and land/use studies relating to National Level and State level physiographic regions are quite essential.
2. Land-form and land-use at the state/district and mandai level with the interpretation of Toposheet as well as aerial photo interpretation is a must and most important and required.
3. Soil/water (surface/underground water, vegetarian, agriculture should be clubbed with socio-economic and infrastructural data to understand the environmental impact and its assessment typology.
4. Patterns of Agriculture/crop diversification, crop producing source agri-infra identified with population growth pattern deficiency/nutritional deficiency and diseases and its impact study is most welcome and is the future areas of research. This study helps the planner to plan health and nutrition programmes of the nation/state and district or micro region, respectively.
5. Urban Morphology/urban spatial relations with rural hinterland / urban poverty is an important study in an administrative region such as Telangana State, Andhra Pradesh and Tamil Nadu.
6. Settlement system in relation to physio-economic, social and Cultural factors - Tribal Prone areas/ flood Prone Areas / Drought Prone Areas/Landslides to be taken up in an administrative region to enable the planner to solve some of the challenges and problems of a region. Bibliometric study has been carried out in this chapter and has been presented analytically in the following paragraphs.

Table 7.1: Indian Geographical Research: Bibliometric Analysis

Sl.No.	Name of the Branch	Single Authored Papers	Double Authored Papers	Multiple Authored papers	Total
1.	Environment & Develpment (B)	0	0	3	
2.	Geographical Landscape	1	0	0	
3.	Geomorphology (B)	1	0	0	
4.	Micro Level Village Study	0	2	0	
5.	Remote Sensin & GIS (B)	0	0	3	
6.	District Level Study—Remote Sensing	1	0	0	
7.	Urban Geography (b)	I	0	0	
8.	Urbanization	0	0	3	
9.	Urban Slums	0	2	0	
10.	Comparative study of Slums	0	2	0	
11.	Rural - Urban Fringe	0	2	0	
12.	Land use & Agriculture Agro Typology District Study	1	0	0	
13.	Kashmir Valley	0	2	0	
14.	Crop-diversification Taluk Study	0	0	3	
15	Livelihoods	1	0	0	
16.	Sustainable Agriculture State Level Study	0	2	0	
17.	Agriculture DPR	0	2	0	
18.	Infrastructural Facilities	0	2	0	
19.	Rural & Region Development	1	0	0	
	District Study (B)				
	Total	7	8	4	19
		(36.84%)	(42.10%)	(21.05%)	

Note: (B) indicates tile papers which were presented in the sr' International Geographical *Congress @ Beijing, China, Asia.*

Analysis and Assessment of Published research papers Analysis and assessment of published research papers are methodologically described in the following paragraphs.

Step- 1 : Nineteen research articles and their titles are presented in table 7.1. All the Nineteen papers who reviewed in this chapter-7 and the papers are first the oretically categorized into three classes depending upon authorship pattern.

1) Single Authored papers, 2) Double Authored papers, (3) Multiple Authored Papers [(2) Refer the table-7.2)].

Step- 2.1 Single Authored papers published in foreign Journals are assigned 100 points.

2.2 Single authored papers published in Indian Journals are assigned 50 points.

2.3 Double Authored papers (Foreign) are given 50 points

2.4 Double Authored papers (Indian) are given 25 points

2.5 Tri pie Authored papers (Foreign) are given 25 points

2.6 Triple Authored papers (Indian) 12 points

Table 7.2: Data Analysis (Chapter-7)

Sl.No.	*Category papers*	*No. of*	*Points*	*(n.p)*	*Total*
1.	Single Authored papers (Foreing)	4	100	4xl00	400
2.	Double Authored papers (India)	0	0	0	0
3.	Triple Authored papers	1	25	lx25	25
4.	Single Authored papers (India)	6	100	6xl00	600
5.	Double Authored papers (India)	8	25	8X25	200
6.	Triple Authored papers (India)	1	12	Ixl2	12
	Total	9	262		1237

Table 7.3: Gradimg of papers based on points score

Sl. No.	*Interval Size*	*No. of papers*	*Grade*
1.	<200	3	C
2.	201-400	1	B
3.	401 and above	1	A

Observations

It is seen from table-7.3, category of papers and their geneses and number of papers that have gained, the points, in the next column (number papers is multiplied by score points is also presented closely with total indices.

1. Single authored papers were to a tune of four which gained 400 points, where as single authored papers were to a tune of six and gained a total points score is 600 points.
2. Double authored papers were published in Indian Journals and their points i.e., 25 has been multiplied, it gives 200 points score.
3. Triple authored papers (India) were one and only one and its assigned value was 12, which yields to 12 points score.

It is seen from table-7.3. grading of papers based on points score has been presented relating to abroad journals as well as 33rd international geographical congress, Beijing, China, Asia, August, 2016.

Observation

(1) Single authored papers have gained momentum, (2) The highest score (401) and above) relates to I with 'A' Grade.

5. Conclusion

1. It is concluded that traditional pattern of urban study is now accompanied by using additional inputs — such as RS & GIS is more attractive and warranted.
2. While studying population concentration, satellite images are used in a number of papers. Besides the studies of sanitation and health under the caption "Geography of Sanitation and Health". This is an essential study and will alterd the Quality of life so in future studies should be conducted.
3. 'Urban Fringe has been divided into three phases. They are: (1) Outer fringe zone,

(2) Inner fringe zone and continuse fringe zone.

This methodology helps to identify and demarcate urban influence in the fringe areas. Urban sprawal studies using climate data which helps to study the heat and besides identify is the spatial variations in temperature & precipitation. Now this branch has been optional as "Urban Sprawl" and its spatial deviation.

4. 'Land-use/and land-cover changes' and 'Urban sprawl' have been attracting may researchers by using RS & GIS methodology helps to devise the categories of land use and its sustainable landcover changes and its influence on human sustainability in a typical urban environments. These are captioned as subfields of geography as "urban environment issues of poverty".
5. Finally, the author concludes that urban zoning and sub-zones should have to be identified for the spatiotemporal variations. 'Urban Poverty' studies may play an important and dynamic role in the planning of employment for unemployed and providing necessary urban infrastructural facilities to alleviate with urban quality. It is a felt need to present micro level planning and Integrated Area Development Studies — A Model, State, District, MandaI and Village level studies. Keeping in view of the above in mind the author has presented the above theme in Chapter Eight which includes research proposal and research paper for inclusive

development for the socio-economic and infra development in an administrative region/Economic region/Zone.

Notes

1. R.B. Singh, Ajay Kumar and Senaul Hagu, *Environment and Resources ", Progress in Indian Geography* - A Country Report, Ed R.B.Singh, 33rd International Geographical congress, Beijing. China, August, 21-25, 2016, pp. 59-76.
2. Hema Thakur, *"Archeological Heritage ill Karnataka with Special Reference to Geographical Setting",* Transaction/ Vol. 39. No. 1, 2017, (pp. 109-114).
3. Viswas, S. Kale, *"Geomorphology",* Progress in Indian Geography - A Country Report, Ed. R.B. Singh, 33rd International Geographi cal Express, Beijing, China, August, 21-25, 2016, pp. 77.91.
4. Ganga Deep Singh Nunda and Subhau Kha, *"Application of Geospatial Technologies for Identification* & Development of Household Database for SCs and Other Back Ward Classes (OBCs) of Sondh Village in *Mewat district* (Haryana)", Transitions, Vol. 39, No. 1, 2017, pp. 69-80.
5. R.B. Singh, Ajay Kumar and Aarli, *"Remote Sensing and GIS", Progress in Indian Geography - A Country Report & R. R. Sing, 33rd international Geographical Congress, Beijing, China, August, 21-25,* 2016, pp. 193, 2013.
6. Chopra Narayan, *"Impact of Human Intervention on Co-Environment - A Case Study using Remote Sensing Techniques", On lime published on* 19 *January, 2017, The Geographer, year, 2016, Vol. 63,* Issue- Z, pp. 1-7.
7. Kalpana Markandey, *"Urban Geography",* 2012-16, Singh R.B (Ed.) Progress in Indian Geography, A Country Report, 20 12-20 16. The 33rd International Geographical Congress. Beijing, China, August, 21-25. 2016, Indian National Science Academy, New Delhi, pp. 105-121.
8. Ashok. K. Dutt & et.als, *"Socio-Economic Correlates of Asian Urbanization"* in Urban Development Volume, Reviewed by K.R. Dikshit. Pune, Transactions, Vol. 39, No.1, 2017, p. 154.
9. Sanyal Srabani & Sen Archana, *"Socio-Economic Conditions of Urban Slum Dwellers in Bhopal City",* online published on 19th January. 2017, The Geographer year 2016. Vol. 63, Issue 2, pp. 61.71.
10. Tripathy, V.K & Wardhan Rahul Harsh, *"Study oj Physical. Demographic and Irfrastructural Characteristics of Slums: Tale of the two Cities"*, online published on 19 January, 2017, The Geographer, year 2016. Vol. 63. issue-2. pp. 81-89.
11. Sadaf and Munir Abdul. *"Spatial Pattern and Differential in Literacy: A Case Study of rural Urban Fringe of Faziabul City",* Online Published on 19 January, 2017, The Geographer, 2016, Vol. 63, Issuez. pp. 133-144.
12. B. Nagabhushanam, N. 2016, *"Land use and Agriculture 2012-16, Singh R.B (Ed) Progress in Indian* Geography : II Country Report. 2012- 16, The 33rd International Geographical congress, Beijing, China (August, 21-25, 2016), Indian National Science Academy, New Delhi, pp. 171-181.

13 Tanveer Ahad and T.A. Kauth, *"Impact of Small Land and Holdings on Rural Economy : A Case Study of Kandi Areas of Kashmir Valley. Jammu and Kashmir",*

Trans Institute of Indian Geographers, Transactions, Vol. 19. No. I. 2017. pp. 105- 108.

14 Siddiqui Shamsul Haque, Asthma, S.P. Akter Narim and Sultana Chand", *Analyzing Spatia-Temporal Pallem of Crop Diversification in 'West Bengal, India, Online Published on* 19 January, 2017.

15 Nagashiu shanam, N, 2016, Ibid - Land usc and Agriculture (128).

16 Abdul Hanan. *"Livelihood. Labour Alurkel and Skill Development in Small Tea Growers (STHs Gardens in India with Special Reference 10 India's North-East".* Transaction, Vol. 239, No.1, 2017, pp. 91-104.

17 Omnir Sigh and Anita, Kurukshetra, Haryana, *"Management of Ground Water Resources for Sustainable Agriculture in Haryana".* Transactions Vol. 39, Vol. I, 2017, pp. 133-150.

18 Ahmed Syed Naushad. Zaidy Md. Naiyer, *"That Slate of Household Food Insecurity in Bundel Khand Region: A Case study of Hamirpur District of Uttar Pradesh".* Online published 19 January, 2017, The Geographer. Year 2016. Vol. 63. Issue. 2. First Page (17) last page (29).

19 Rahman Fazlur, Ansari Azimulluh. *"Changing Focus of Health Care Delivery System in India : An Evaluative Study* Online Published on 19 January, 2017. The Geographer, year 2016, Vol. 63, Issue 2, First Page (104) last page (113).

20 Mohan, Krishna, 2016, *"Rum! and Regional Development, 2012- 16",* Singh R.B. (ed.) Progress in Indian Geography A Country Report, 2012- 16, The 33rd International Geographical Congress, Beijing, China (August 2 1-25, 2016), Indian National Science Academy, New Delhi pp. 161-170.

Chapter - 8

Micro Level Planning and Intergrated Area Development - State & District Level Studies

An attempt has been made by the author to present the micro level planning and integrated area development studies: A model in particular to the state, district and village in a mandal study in this chapter. This study recognizes and identifies the following aspects. They are: (1) Statement of the problem, (2) Survey of research literature and gaps in literature, (3) nature and scope of the study, (4) Objectives and Hypothesis, (5) Methodology, (6) Tentative chapterization, (7) Bibliometric methodology, (8) References, (9) University Centre and district headquarters centre and its role in the development of geography, (10) Research papers & Research proposals.

1. Statement of problem

In the political landscape of India, 2014 is an important year as TPS government has assumed office by the present first Chief Minister, Sri. K. Chandra Sekhar Rao (KCR) for Telangana State, which is the state of the Indian Union. In the annals of the geopolitical history of Andhra Pradesh 1982 is an important year as Telugu Desam Party (TDP) government has assumed office by the then Chief Minister, Shri N.T.Rama Rao, who has introduced a series of administrative reforms. Foremost among them was the introduction of 'Mandal' system in the place of 'Block' system. Prior to this block (Micro) level consisted at the lowest and also below the district and state administration respectively. In order to capture the extent to which this mandal process, there is a need to build in incentives that will encourage status to devolve functions through 'spatial planning'.

The present study is an attempt to examine the 'Mandal' system is helping inclusive and integrated development in the spatial units over a period of time (programmes and performance appraisal). In this process, it goes deep into the secondary data on population, agriculture, industry, infrastructure, Urban structures and allied facilities to facilitate the existing population and its impact of the system in development through spatial planning at micro level. The study also analyses village survey data to capture the main planning effects through this reform (evolution of mandal system) in areas of economic and social development with an index of development i.e. physical index, economic index and infra index for the study region, mandal which helps for achieving the balanced development and also the further need to initiate/constitute or plan and design new districts in the newly formed 20th state (i.e.) Telangana state in order to facilitate and access to important functions or needs of the people which have to be translated through spatial planning, and finally achieve balanced and integrated (area) development.

Keeping in view of this in mind, micro level planning and integrated area development - model is presented for conducting an indepth and detailed study in order to suggest the planning of growth centres/service centre planning strategy, so as to achieve and fulfill the objectives and also to wipeout regional and spatial inequalities at the micro level and also designing of new district(s) will be carved out from the present and existing largest (areas) districts into smaller (micro) administrative units/areas for administrative viability through micro level planning and integrated area development strategy which evolves to involve people in the entire planning and development process for human sustainability in a given and identified environmental scenario.

2. Survey of Literature

2.1 Regional Planning has become a part and parcel of modern national planning and has created new place for contributive research in,geography. It deals with the delineation of hierarchy of regions, based on the physical, human resources, and an appraisal for a balanced and integrated area development.

The regional approach and regional problems were identified and recognized in the first and second five year plans for development though not much was done to bring them to the desired level effectively. In the second five year plan, district plans were prepared taking the territory as basis but not much was done here also. Many of the district boundaries cut across physical and agronomic region and preparation of state plans and block development programmes included in the third five year plan for rural development were the incipient approaches to planning.

2,2 Hierarchy of planning regions - Concepts

Planning should be limited to the degree of economic viability and the scale of major development in the area. The 'area' within which various development programmes could be organized can be classified as macro, meso and micro. Thus, it is possible to classify planning regions of various taxonomic ranks by combining area of homogeneity according to the purpose and scale. Macro region's boundaries may cross from one district to the other or even within the state also and sometimes may hamper the economic aspirations of the people and micro regions may have similar characteristics among the macro region, which enables the planner to identify and study the resources indicates for the judicious utilization of resources and the planner invariably faces some problems of its own. Micro regions should satisfy some of the basic needs and desires of the people. Structural, functional and organizational aspects should be viable for direct interaction with the service centres and also with the growth centres.

The heterogeneous composition of people and regions and wide regional variations in their development indicates that a plan should bc based on local resource potential. The necessity of area based micro level plans for correcting this situation was recognized in the fourth five year plan itself. The planning commission has issued a brochure in 1969 (the first year of the fourth five year plan) setting-forth of a set of introductory guidelines for the preparation of plans. The emphasis was much on data collection than on analylitical techniques.

2.3 Spatial Planning

'Spatial Planning' is necessary to achieve spatial development at diffident area levels such as Nation, State, District and Other lower administrative units. Three levels of regional hierarchies are important and there is every need to plan for micro level units in order to achieve development. The structure, process and stage elements are important in spatial planning methodology. This is essentially the study relating to resource utilization and their spatial distribution. Geomorphology and climatology have a bearing upon the regional physical structure besides socio-economic and cultural composition. The transformation technique which takes its full form if we provide some more socioeconomic facilities in a region/mandal to upgrade, step-up or update them in order to meet the needs of the growing population to achieve balanced and integrated development in its full form with structural planning.

It is highly desired to identify and recognize variables which causes variations in the levels of development. To sort out the indicative variables, some of the basic facilities would be provided to achieve regional development or spatial development over a period of time. The basic amenities will certainly alleviate the quality of life of people living in genetically problem oriented areas such as flood prone areas, agriculture prone areas, drought prone areas, and urban prone areas and so on. The scenario of adjoining areas of un-irrigated agricultural systems (urban shadow regions) and tribal prone areas are also important to design a spatial framework for development that should include plan for land use, human settlements, social services and amenities that are required by the people. In order to bring development at the grass-roots level, planning will have necessarily started from the bottom instead of the top.

If the micro level units are properly planned, they help in the maximum utilization of resources for development. Further, these potential units help to support the rural poor at the micro level. This inturn curtails rural migration and congestion which reflects in urban areas. In view of this situation, micro level plans can help in fixing priorities for different regions depending on their specific needs, without - micro level planning, no national plan could properly be augmented.

The investigator plans to present the following aspects for critical evaluation which enable to present current research profile in the following paragraphs. They are:

1. Statement of the problem
2. Survey of literature
3. Nature and scope of the study
4. Objectives
5. Methodology
6. Tentative Chapterization; and
7. Bibliography.

The research Iiterature is presented into two phases. They are: I) Before 2000 year and 2) After 2000 year.

Major contributions that have been made by Indian geographers are quite important and the ideas were portrayed through Christaller's central place theory.

Bhat L.S., (1976)[1] in his study on "Micro level planning - A Case study of Karnal. Haryana, stressed the need for micro level planning in India for minimizing the regional imbalances and proposed the use of growth centre techniques.

Prakasa Rao, V.L.S. (1977)[2] in his study on "Development Strategies for an agricultural region - A case study of Muzaffar Nagar District, U.P'· analytically dealt the intimate relationship between spatial structure of rural economy and the speed and pattern of its development with a focus on Muzaffernagar district of Uttar Pradesh, he suggests a spatial strategy for rural development consistent with the modified central place theory of Christaller. In addition to the earlier empirical findings of Christaller, Rao has identified the factors which promote or retard growth/development in a predominant agricultural district. Grap deactivate factor analysis and correlation co-efficient were used to delineate the planning regions and suggest a workable plan for the agricultural region.

Secondly, the classification method of regionalization is now based on various modern procedures. The derivatives of the above discussed multifactor uniform regional patterns based on synthetic approaches of regionalization have now been processed logically by applying "factor analysis to workout the complex and co-associated nature of areal properties 'linkage analysis' . Such processes have been intensively applied by disaster management and spatial-frame-work. geographers and planners Berry (1961), Berry and Rao, V.L.S.P (1968), Pal (1963) etc. for regionalization and planning[3].

Reddy, Y.V and Kosalaram, K. (1988)[4] have presented 'a study of Planning with specific focus on the considered practices and perceptions with regard to multi-level planning as a whole in India. The theory and practice of multi level planning with special reference to India and also listed a number of emerging issues with regard to multilevel planning in general and decentralized planning at sub-state level in particular.

Sectoral and spatial planning covers various sectors and includes plan formulation. M.S. Swaminathan has presented ,"Alleviation of Social and Regional inequalities and highlighted the aspects such as, the removal of social and regional inequalities and economic growth comprise the two objectives of planned development. Plans and programmes are in operation with the objective of removal of both social and regional inequalities. The social inequalities are poverty, unemployment and poor quality of life. These are sought to be combated through three types of programmes viz.

(a) endowing income generating assets on the poor, (b) employment generating programmes and (c) provision of direct consumption assistance to specially vulnerable sections of population under all these programmes as under all other developed activities, defined proportions of benefit are to flow to scheduled castes scheduled tribes and other backward classes of the society at different area levels. The combating of social and regional inequalities was given first place. "An important element of the policy of the present government is to get into contact with all the poor people in the state in terms of provision of employment and also strengthening the social service facilities in terms of water supply, sanitation, housing, medical, education. Rs. 2Kg Rice, mid day meals etc.

Programmes have to be designed to ensure long term development of those areas such as genetically drought prone areas and flood prone areas etc. An objective of reduction in regional imbalances in the level of economic development also would have to be "identified"

The district planning process is expected to ensure accelerated development of backward areas. However, in order to enable the district planning process, to bring about. such accelerated development activity, the allocation to various districts will have to take into account the levels of development and the distribution of state level activities should take into account the disparities in the development of the district.

Allocations are dependent on the following factors: (a) Population, (b) S.C Population, (c) S.T. Population, (d) Geographic area, (e) Length of sea Coast. (f) E.P.A (affected area) Flood prone areas, (g) Gaps in the social infrastructural facilities in the context of the establishment of mandals and needs of supplementary employment programmes.

Mohammad, Afzal (1988)[5] has presented, "Spatial Framework for Development Administration and Planning and Suggested an Innovative research methodology for identifying growth centres, service centres and central villages by adopting ranking method.

Post 2000 year

Remarkable research contributions were made by scholars during the period of post 2000 year may be classified as quantative revolution in research methodology in geography, statistical methods were used in the analysis part of research and quantitative research methodology, which helps to arrive at with precise calculations leads to valid conclusions.

Regional inequality in Odisha: Some emerging issues were analysed through factor analysis by Mohdizhar Hussain (2007)[6], L.T. Nayak[7] has presented Regional disparities in levels of development in Bellary district - Karnataka through composite index method (2007) and G.R. Diwan[8] has presented urban and regional planning activity in Maharastra using both qualitative and quantitative research methodology (2009). These three studies are important contributions to the modern geographic research and also help to democratic planning regions.

'Agriculture' is the chief aspect of development which has to be studied in terms of broad landuse and cropping pattern at different area levels with regard to agricultural typology and agricultural geographic research activity have been stemmed into the following forms. Naushad Ahmed[9] has presented Regional imbalances of food security in, the Northern Bihar Plain (2007), Nizamuddin Khan[10] has presented 'Regional spatial patterns of agricultural development in Murshirabad district, West Bengal (2007), Mohd Julfikar Ali[11] has presented, Micro Regional and geographical analysis of Agricultural crop productivity (2006) are important contributions in the field of regional agricultural geography and typology, especially helps to identify spatial patterns of agricultural development besides spatial variability.

'Population' is a source and resource and it is one of the aspects of spatial development which has to be studied in depth and detail. With regard to the population analysis, the special mention may be made to D.H. Power's[12] paper, 'Geographic Analysis of population pressure on land in Maharastra and regional disparities of levels of Human Resource development in Southern Plateau Region of Maharashtra by C.T. Power[13] are really an important additional contributions to demographic studies besides, population mappmg.

'Irrigation' is a powerful tool to change the economic structure with market centres and its hierarchy in a region. Economic activity is presented in different geographic regions by many geographers. Among them, the scholarly contributions are as follows. Shashid Imam[14] has presented, 'Economic Impact of Indira Gandhi Canal project, Western Rajasthan, C.T. Power[15] has also presented a theme on, 'Role of market Centres in Regional Development: A micro level Analysis' (2004), Manas Pal1, presented thematically on, 'Response of Market surplus: A case study of paddy in Burdwan district.

West Bengal'(2009) and N. Das[17] has presented an exhaustive analysis entitled, Hierarchy of Market centres in the Rural Hinterland of Cuttack - Bhubaneswar, Orissa, India. The above studies are important contributions to agriculture, regional disparities, market units hierarchy and also helps to understand the regional patterns of development in agricultural and allied activities and also further to understand the logic of market centres and its interaction with the economic units in each and every activity to the total economic activity in its entirety.

Regional geography, regional analysis, regional resources and regional development may be identified through factor analysis and it is an important tool for geographic analysis and helps to understand the variations quantitatively in a region.

Jay Sree, D[18] and et.als have studied the disparities in the levels of development in the districts of Gujarat, which pin point to understand well the factors that have contributed to the differential growth patterns. Their study has identified industrial activity factor as the major determinant of the spatial variations in the overall economic development. The author has presented 19×18 data matrix explained the three factors which together account for 80.5 percent of the variable (explained variable) and its variability.

Ahsis Sarkar'[19] and Sandip Das studied the agricultural and resources of Hoogly District, West Bengal. Being located over the delta plain of the Ganges, land resources of the Hoogly district are common and self-rich agriculturally with regard to the soil. The spatially new uniform geo environmental parameters and the varying human landscape together influence the existing land-use pattern, making it typical multi faceted problem.

S.C. Shukla and S.Dhagat (1996)[20] studied the patterns of land utilization in Madhya Pradesh: A Factor Analysis Approach was presented and high-tightened the patterns of land utilization in Madhya Pradesh.

T.N.A. Rao[21] has presented "Balanced Regional Development: A Development plan for Karnataka and presented the factor scores and correlation matrix which would pin point the positive and negative aspects of development of 175 taluks based on 39 selected indicators. Further, opined the study indicates that political factor play an important role in regional geographic resource development. The 73rd Constitutional Amendment has provided a statutory power to the local government to undertake micro level planning and development exercise. This work is an invaluable source of basic statistics for regional development planning. Keeping in view of the structural research literature, some important gaps are identified and important aspects are to be taken up for the present research problem.

Gaps in Literature

1. There is no study relating to the identification of heterogeneous factors in a homogeneous plain plus irrigational sources that have perennial indelible mark on development. In the backward region of Telangana state a forward district with less heterogeneous characteristics may be identified by making use of Factor Analysis Technique and also ranking method in order to achieve the levels of development in the spatial units.
2. There is no single study relating to mandai or a district in Telangana State, India stressing the Regional development taking a case study for development planning micro level planning is a necessary planning tool for assessing the resources and plan for development. Hence, the present study is related to (ie) " Issues and challenges of Micro level Planning for a District, Telangana State, India.

3. Nature and Scope of the Study

The study has its significance at large by covering geographical, economic, social and infra factors in the regional economy particularly mandal & district administrative machinery. To overcome the future exodus of population (exponential growth rate) some viable and possible measures are to be suggested for the mandal and district administration in order to achieve an optimum and balanced status besides bringing people above the poverty line. Keeping in view of the various elements and factors associated with the mandai and district administration it helps the top district administration to take-up the viable steps for incorporating some of the suggestions as policy decisions for alleviating the poverty of the people and make them to step-up the quality of life in the micro administrative unit. Keeping in view of the socio-economic situation of the people at the micro level, it is highly desirable to adopt ways and means to achieve the inclusiveness as well as 'integrate' development at the district level covering the uncovered aspects'so far.

The principal aim should be development of semi-developed areas covering the villages of the mandai, which in turn helps to economic and social development through micro level planning as a strategy for spatial development planning. Further, keeping in view of the identified problems and find solutions for the same through this spatial planning exercise would prove to meet the objectives of planning and development within a given period of time. It is aimed at the development of 'mandals' and finding solutions for the problematic areas of the sub-unit level and also at the village level. The roposed plan helps for the future population of the district in general and cluster of villages in particular. It not only solves the present day problems but also the very basic needs of the society in future.

4.1 Objectives of the study

Objective-I: To examine and study the trend pattern of population growth at the, micro level (ie) mandal level (2001-2011) and estimate the future population growth for the year 2021 (Growth Dynamics) and also to plan the infrastructural facilities for inclusive growth and integrated development. The settlement pattern of the villages are expected to expand beyond the settlement boundary, thereby containing population growth therein static terms of landuse and population change, some stipulations have to be prepared to the urban growth.

1. To prepare population growth for all the villages of the mandal should be studied in detail based on the secondary data, which is available in Hand Book of Statistics,
2. To prepare the upward mobility immigration and out migration of population and their size,
3. Population size: non factor in population growth,
4. 'Distance' as a function of variability and non visibility of urban influence and the absence of urban infrastructure. A factor in population growth (0-5, 5-10 and 10-15 KM) radius and drawing of concentrate circles and calculate the functionality as well as population threshold which helps to identify the felt need of infra structural facilities so as to understand the adequacy or inadequacy of facilities and functions.
5. To study the influence of national Highway and adjacent to other states.
6. To study the Out-migration patterns and influence (peripheral areas) and future scenario.
7. "Increasing distance' leads to a decrease in central functions. Third order, second order and first order centres are to be identified from the mandal/district head quarters and also to examine the transport network and its influence over space and people.
8. Linear, compound and exponential growth if any should be identified by population size category of units (population groups).

Objective-2: To prepare a systematic documentation and analysis of events that motivated city development of immigration in general and district in particular.

In order to reach the second objective, it is a necessary prerequisite to prepare a systematic documentation and analysis of event that motivated development of immigration in general and the selected district in particular. In any district, Karimnagar district, Sriramsagar was built and the phase wise excavation of canals determined the phases of development of migration in regions such as genetic hierarchic agro regions of the district I) main portion / sub portion and tail end (dry region). Analysis of impact of irrigation in terms of gross sown area, cropping intensity, extent of irrigation, type of irrigation and cropping pattern data has to be collected for analysis. (Time series data will be collected and analyzed. 2) However, the gap is for identifying the patterns of changes eliminated during the post construction period due to the commencement of new irrigation projects in moderate and dry regions that were sponsored by the state. This objective identifies and stresses that with a proactive role of the state by focusing on less developed regions with appropriate policies that Could successfully bring down spatial inequalities across the region.

(1) Genesis of irrigation system through Godavary river waters.

(2) Irrigated area in the district / length of the canal in KM. (first crop/ second crop/mandals benefitted.

(3). Extension of canals to the tail region of the district.

(4) Irrigated area through minor (irrigation tanks and lift irrigation (2001-2011) rates to minor tanks/lift irrigation to understand the irrigation sources and associates

(5) Agriculture and spatial inequalities (measures over a period of time).

(6) Percentage distribution and simple growth rates of GSA by broad in a district (spatial temporal variations).

(7) Patterns of intensity at Macro level and percentage distribution of GIA (Spatiotemporal), type of irrigation, sources of irrigation and cropping pattern etc.

Objective-3: To identify the Central place Hierarchy such as growth centres/Service Centres/Central villages through ranking method, according to the centrality score of the mandals/settlements based on Christaller's central place theory.

At the first level of functional hierarchy, the following functions are to be considered.

1. Primary school
2. Centre for Medical Checkup and Preventive medicine
3. Branch Post Office
4. Request Bus stop
5. Regular Bus stop
6. Primary Credit society
7. Retail daily market
8. Seeds distribution centre
9. Fertilizer distribution centre
10. Agricultural implements distribution centre

These are the policy functions and the remaining are non-policy functions.

First level (Population 10,000)

Second level (Population size 10,000-20,000)

1. Middle School (UP)
2. Secondary School (SSC)
3. Allopathic treatment and Hospitalization centre
4. Sub-Post office
5. Bus Station
6. Scheduled Commercial Bank
7. Animal Husbandry, Supervisory Unit
8. Pesticides and Fertilizer Depots.
9. Hardwar Shops
10. General Provision stores and
11. Cloth stores.

Third level (Population Size above 30,000)

1. Junior college
2. Centre for surgery and Hospitalization
3. Centre for Maternity Care
4. Centre for Family Welfare Schemes
5. Head Post Office and Telegraph Office

6. Telephone Exchange
7. Bus Junction
8. Land Mortgage Bank
9. Veterinary Hospital
10. Restaurants
11. Chemist and Druggists Shops
12. Whole sale and Regulated Markets
13. Jewellers Shops and
14. Glasswork and Potters

Service Centre pattern (Observation) and identification based on ranking method. This analysis helps to understand the process of growth centres, Service centres and central villages and the planner may be able to plan the new service centres which improve the travel pattern, travel cost pattern and travel time of the consumer. Consumer(travel) to satisfy the needs of persons to reach the service centre (space preferences). The new service centre planning will facilitate the population for interaction from village to the higher service centre to satisfy the needs of the people along with transportation method (existing).

Objective-4: To identify the regional inequity in the study area and also to study the economic and social change (scenario) in order to assess the levels of development.

To achieve the above objective, there is every need to identify, recognize and assess the economy for social development and direction of change in regional inequality during the study period, the factor analysis method has to be applied in order to identify the levels of development broadly. This technique may be applied for assessing the performance and role of variables in attaining the aggregated and disaggregated picture which helps to understand the spatial inequality and also further to find sources to achieve equality (overall).

Objective-5: To identify the allied agricultural activities such as poultry farming and milch cattle pogramme (white revolution) and its influence.

This study concentrates on the spatial distributional activity of the above have to be assessed through the collection of primary data besides secondary data and also map them.

Objective-6: Collective strength is more stronger than the individual strength/influence of self help groups.

To identify the economic and social development process in the study area, such as micro-entrepreneurship with Self-Help Groups to achieve economic development, both primary and secondary data have to be collected in order to assess the level of development over a period of time. To evaluate the process, the primary data have to be collected at micro level (village) units and plan future development for the growing population and the areas for the ongoing process of development which may help to incorporate it in the district planning process.

4.2 Hypotheses

1. The district is uniformly developed,
2. Structural, functional and organizational factors are viable for direct interaction with the service centre,
3. There is no need to inject infra facilities to match the population growth for the balanced and integrated development of the mandal/district.
4. If the micro level units are properly planned, they help to the maximum utilization of resources for integrated and balanced development of the study area.
5. Travel time/travel cost and travel distance may increase in order to reach the service centre by the consumer (traveler) from the lower level units or group of units (clustering of units).
6. Clustering of functions may proportionately decrease with the increasing distance from the growth/service centre where in people, goods and services interacts in a geographic region.

5. Methodology

Theoretical Framework

The Central place theory provides a framework for understanding the regional spatial structures and in regional planning on hierarchical system of centres avoids duplication and waste. The application of the theory in the regional planning of some of the developing countries like Ghana (Grover and Huzer, 1964) and Israel (Jackson, 1970) have proved to be useful.

The basic idea of the theory is that economic activity within a region tends to agglomerate around a small number of focal points. Polarization flows will gravitate within a region towards these focal points with a gradual reduction in the flow density by distance with the help of minimum critical level of flow densities, a boundary may be demarcated around a focal point which can be described as the growth point., On this interpretation the spatial distribution of population can be regarded as being organized into hierarchical system of nodes and functional linkages. "Stronger the nodal characteristics of the regions higher their growth rates and their levels of economic and social development are likely to be. Regional plans will be more successful if they effectively reinforce the natural nodal characteristics already in evidence in a region.

Identification of areas having different levels of economic development is an important aspect of regional development studies have been conducted in this area using district and state level data for different parts of the Country and India. Regions have also been identified on the basis of their functional linkages revealed by the commodity flow. Topical studies include works dealing with some of the aspects of development individually or in combination with others and policies and framework for urban and regional planning.

Industrialization is closely associated with high economic development location of industries is, therefore, an important factor in the process of industrialization. Location of specific industries with the available sources in relation to district and regional development perspectives have to be analysed. The trend in general has been to review the present industrial structure, problems of growth and prospects for different regions mostly in qualitative manner.

Urban surveys have to be conducted at various levels for tackling problems of development through identification of least urbanized areas and regional imbalances in urban infrastructure. Integrated rural-urban development approaches have to be followed in this particular study.

Urban centres and rural communities both play crucial roles in the development and urban rural process. The whole spatial human settlements villages, movement towns, intermediate cities and metropolises must be strengthened and integrated into well articulated spatial system performing specialized and diversified production, distribution, consumption and exchange functions.

It is thus helpful in reinforcing a network of development centres which can counter balance the pull of urban centres. Through these centres it is possible to transform the subsistence agricultural and rural sector into employment and income generating elements of National economy. The hierarchy of urban settlements was attempted and tested against the postulates of Christaller's central place theory.

2. Factor Analysis Method

Sectors and development indicators are spread over into four groups. They are:

1. Agriculture
2. Economic infrastructure,
3. Social infrastructure and
4. Population

Under agriculture, five factors have to be selected for identifying the variations broadly. Likewise, under economic infrastructure, 6 factors have to be selected for analysis, under social infrastructure, four variables have to be selected for analysis. Finally under population, six factors have to be selected in order to identify and recognize variations at the micro level. The factor analysis method could be applied for understanding the variations in the macro as well as sub regional levels. Keeping in view of the above basics, the investigator should aim and prepare service centre planning for achieving the objectives through a well defined plan.

Sectors and Development Indicators:

Agriculture

AI =% of Gross Area Sown to Net Area Sown.

A2 =% of Food Crops to Gross Area Sown.

A3 = % of Area under Commercial Crops to Gross Area Sown.

A4 = % of Net Area Irrigated to Net Area Sown.

A5 = % Livestock Units per 100000 Population.

Economic Infrastructure

E1= % No.of Post Offices per 100000 Population.

E2 =% No.of Tele Telephone per 100000 Population.

E3 =% No.of Motored Two Wheelers per 100000 Population.

E4 = % No.of Co-operative Societies per 100000 Population.

E5 = % No.of Bank Branches per 100000 Population.

E6 = % No.of Enterprises per 100000 Population.

Social Infrastructure

SI = %Villages having medical facility.

S2 = % No.of Doctors per 100000 Population.

S3 = % No.of Beds per 100000 Population.

S4 = Teacher - Student Ration.

Population

PI = Sex Ratio

P2 = % of Urban Population to Total Population.

P3 = of SC and ST Population to the Total Population.

P4 = % Agricultural Workers to the Total Workers.

P5 = % Industrial Workers to Total Workers.

P6 = Total Literacy Rate.

5.1 Nature, level of study and methodology

Keeping in view of the above scenario in mind, the present study aims to examine and evaluate the performance as compared to the overall economic and social development process in the district planning relating to two census years. (2011 - 2021) as it further examines the trends in regional inequality in the study area during the study period and also to wipe out the inequalities in order to achieve integrated development in the study area, a development plan will be prepared and presented for its effective implementation.

This study is based on performance of indicators both at the district as' well as 'mandal' level drawn from various secondary sources basically concentrating on 'focus group' as a resource to collect the data for further processing and also to identify a set pattern.

The secondary sources like statistical abstract, District census Hand Books, Economic survey and census publications are helpful specially to identify spatial relations relating to 'area' and 'population', 'Area' and 'population are the two deciding factors to identify the need and recognize as suggestive norm for locating the facilities, thus thereby achieve the balanced development. The performance of economy at two levels, to be examined as one related to the district at Macro level and mandals performance as meso level unit reveals the regional inequality in the district present. The investigator has to identify the inequality among the mandals to identify the inter and intra mandal variations in the levels of development in the sectors such has agriculture, economic infra, social infra and social development process is based on population as well as area.

Agriculture, socio-economic and cultural variables are identified as causative factors for regional development. Population has been taken up for analysis at two levels. The first area is related to mandal level (micro level) and the second one is related to the district level (Macro level) studies may help to identify the regional inequality over a period of time. The sector and development indicators are considered for anlaysis in order to assess and arrive at the status of the mandal's performance in the study area,' varying from macro level to Meso level. The sectoral and development indicators are to be presented relating to agriculture, economic infra structure, social infrastructure and population helps to further processing and carrying out ' factor analysis' method.

5.2 Research Design (Policy decisions)

The creation of new districts are entirely based on the act of 1974 of Andhra Pradesh. This act has come into force from 15th February, 1974 creation of mandals has been introduced by the Act of 1985. The questions relating to the formation of the districts are:

1. Geographic area and extent.
2. Population of the district.

Keeping in view of the two points in mind, an additional point is to be considered (ie.) assembly segments. Generally, the assembly segment and its extent solely depend on the district alone. This is a major and nodal concept while forming a new district.

29th Telengana state has been formed in the Indian Union with ten districts. They are: (I) Adilabad, (2) Nizamabad, (3) Karimnagar, (4) Medak, (5) Hyderabad, (6) Ranga Reddy, (7) Mahabubnagar, (8) Nalgonda, (9) Warangal and (10) Khammam. The population of the above districts are as follows:

(1) Adilabad (27,41 ,239), (2) Nizamabad (25,51 ,335), (3) Karimnagar (37,76,269), (4) Medak 30,33,288), (5) Hyderabad (39,43,323), (6) Ranga Reddy (52,96,741), (7) Mahabubnagar (40,53,028), (8) Nalgonda (34,88,809). (9) Warangal (35,12,576) and (10) Khammam (27,97,370).

While creating new districts, these aspects have to be necessarily considered as "basics'. They are:

I. How many districts have to be formed?
2. What are the district head quarters?
3. In which districts the assembly segments are to be formed?
4. New districts geographical boundaries?
5. The towns that are suitable for district head quarters with.

6. The existing infrastructural facilities
7. The felt needs are to be identified to facilitate as district head quarters.

Eenadu, Thursday, September, 10, 2015 (p.ll). Model District for study (an example)

Telangana state: Total No.of Districts

SI,No	*Name of the District*	*SI,Nc*	*Name of the District*
1.	Adilibad	16.	Khammam
2.	Manchiryala	17.	Kothagudem
3.	Nirmal	18.	Medak
4.	Asifabad	19.	Sanga Reddy
5.	Karimnagar	20.	Siddipet
6.	Jagityala	21.	Palamur
7.	Paddapalli	22.	Wanaparthi
8.	Sircilla	23.	Nagarkurnool
9.	Nizamabad	24.	Gadwal
10.	Kama Reddy	25.	Nalgonda
11.	Warangal (Urban)	26.	Suryapet
12.	Warangal (Rural)	27.	Yadagiri
13.	Bhoopalapalli	28.	Vikarabad
14.	Jangarrus"	29.	Medchal
15.	Mahaboobabad	30	Ranga Reddy
		31.	Hyderabad Urban

As per 371D, the Government of Telangana, the problem of Zonal System (Presidential Order) 1975, Local category of candidates has to be recognized by the Supreme Court of India. It is a challenge to Government of Telangana, to face the filling of the posts (Employment) by the qualified candidates keeping in view of the prevailing zonal system of the formed Andhra Pradesh State in vogue.

Karimnagar district has 57 mandals covering an area of 11,823 KM_2 with a population of 37.76 lakhs and the factors relating to physical, socio-economic and infra factors which contribute for the overall development. That is to say that 57 mandals have been spread over, data relating to the districts are segmented and divided into three Natural Regions, (1) The Northern Region, (2) The Central Region and (3) Southern Region.

The study area Telangana state is a classic example of a geographical region which stands out as a less developed sub-region. It also emerges out less developed part of the Deccan Plateau Region comprising parts of

Maharastra, Karnataka and Andhra Pradesh.The region has marked Contrast in the regional structure and processes of development. The dichotomy between markedly rural and urban population with limited agriculture on the one hand and the locations of basic and heavy industries which are by and large the only centres of economic importance at the national level, on the other hand, is quite apparent. The spatial processes of development have thus a marked discontinuity leading to low levels of development in spite of heavy impacts on major industries. The space relations are relatively weak among the modal settlements in term while the regions external relationship through the major industrial centres is quite strong". It is against this background that salient features of selected concepts and theories related to regional development are reviewed in the fore-going pages. What follows in the study of Telangana district of state is an attempt to portray the spatial processes that are at work within the region which serves to bring-out the causes for under development of the region and a way forward will be presented.

Present Study

Keeping in view of the scenario, objectives and methodology in mind, the present study aims to examine and investigate the performance with the bare (basic changes) comparable to overall economic and social development process of the district relating to census years (2011-2021). Further, it examines the trends in regional inequality in the study area during the study period and also to wipe out inequalities, an integrated development plan would be suggested for its implementation and its augmentation.

The present study is based on both the district in general and Mandal Pattern' in particular, drawn from various secondary sources besides primary source. The secondary sources like statistical abstract, district census Hand Book, Economic Survey and other census publications, specially relating to spatial relation and also to study the intra and inter spatial relations relating to 'area' and 'population' .

Plan Period and Sample for analysis

The plan period proposed in this study covers a period of 10 years (ie) from 2011-2021.

6. Tentative Chapaterization

Chapter-1: Introduction

Chapter-2: Profile of study area

Chapter-3: Resource Appraisal of the study area - Geography - Economics and Population of the Area.

Chapter-4: Agricultural Resources, Agricultural Allied Resources & industrial Resources of the Study Area with a Micro Appraisal and also the Social Structural Developments

Chapter-5: Central Place Theory and its Application in the Study Area.

Chapter-6: Factorial Analysis - Process of Decomposition

Chatper-7: Factor Analysis (Theoretical Formulations and Conceptual Geographical Variables - A Factoral Technique) (Non-Mathematical Approach) - An Innovative Theory.

Chapter-8: Identification of Regional Disparities (Area Specific and Population Specific).

Chapter-9: Strategy for Integrated Area Development.

Bibliometric Analysis:

Bibliometric analysis has been presented in the following paragraphs in order to identify and recognize the authorship pattern in the areas of Micro Level Planning and Intgrated area development The core study and concent-ration upon different aspects of region/regional planning and regional development in this chaprer are:

Table 8.1

Sl. No.	*Title of the paper*	*Single Authored papers*	*Double Authored Papers*	*Three Authored Papers*	*Total No.of Papers*
1	*2*	*3*	*4*	*5*	*6*
1.	Micro Level Planning	1	0	0	
2.	Development Strategy	1	0	0	
3.	Regionalization & Rural Planning	0	2	0	
4.	Micro Level Planning	0	2	0	
5.	Development Administration & Planning	1	0	0	
6.	Regional Inequality	1	0	0	
7.	Regional; Disparities in Levels of Development	1	0	0	
8.	Regional Planning Activity in Maharastra, India	1	0	0	
9.	Regional Imbalances of food security	1	0	0	

(Contd...)

10.	Regional spatial Patterns of Agricultural Development	1	0	0	
11.	Micro Regional and Geographical Analysis of Agri-Crop productivity	1	0	0	
12.	Geographic Analysis of Population Pressure on land in Maharastra, India	1	0	0	
13.	Regional Disparities of Human Resource Development	1	0	0	
14.	Economic Impact of Indira Gandhi Canal Project to Rajasthan	1	0	0	
15.	Role of Market Centres Regional Development	1	0	0	
16.	Response of Market Surplus - Burdwan District, W.B.	1	0	0	
17.	Hierarchy of Market Centres - Cuttack - Bhubaneswar	1	0	0	
18.	Disparities in the levels of Development in the Districts Gujarat	1	0	0	
19.	Agriculturae. and Resources of Hoogly district, W.B., India	0	2	0	
20.	Pattern of land Utilization Madhya Pradesh	0	2	0	
21.	Balanced Regional Development - Karnataka state, India	1	0	0	
	Total	**17**	**4**	**0**	**21**

Table 8.2: Assessment and Analysis

Sl.No.	*Category*	*No.of Papers*	*Points*	*Nep*	*Total*
1.	Single Authored Papers	17	50	17×50	850
2.	Double Authored Papers	4	25	4×25	100
3.	Triple Authored Papers	0	0	0	0
	Total	**21**	**75**	**0**	**950**

Table 8.3: Grading of Papers

		Grade
Below100	1	B
100-500	0	C
Above 500	1	A

7. Applied Bibiometric Methodology

It is seen from table 8.1, that the detailed pattern of titled papers with authorship typology. A total number of 21 papers are presented with their captions. In the third column, to 6th column, single authored papers, double authored papers and three authored papers particulars are presented to facilitate the reader/researcher.

Analysis Observations

1. It is seen that there are 17 single authored papers out of 21 papers and they are really good contribution to current research.
2. Four papers only out of 21 papers, forms as double authored papers. These papers are related to technology and other statistical treatment papers assessment and analysis has been presented in table 8.2.

1. It is seen from table 8.2, 17 single authored papers takes the top position and each paper carries a weight of 50 points, then 17 has been multiplied by 50 fives 850 point score.
2. It is seen from table 8.2, 4 double authored papers take the second place out of the total sample of 21, and 4 has been multiplied by 25, it gives 100 score.
3. In the entire sample, there are no trible authored papers, so there are no scored points'f'
4. The total score for single authored are 850 and the total score for double authored are 100.

In table 8.3, Grading of papers are presented. Finally inclusive of 100, one category double authored takes the 'B' grade where as,500 and above score, the Singe authored papers takes the lead with 850 score and second 'A' grade in the total study of the current research.

This methodology helps to recognize and identify the authorship pattern and also for grading the papers according to its genesis besides the pattern of change in authorship suggests for future publication and management process thereafter.

Part-b

Research Proposal

Economic and social study of Trade, Transport, Market Systems and Rural Development Processes: A case study from Andhra Pradesh, India.

Introduction

Economic and social activity over space is unlimited but it varies widely and vividly in the economic and social sector. Physical environment play an important and vital role in the economy whether it is agriculture, industry, trade, commerce and industry related development through a net work of systems of transport and allied phenomena. The principal aim of this research examines and identifies the economic activity over the earth surface and it's an inter-related phenomena which accelerates employment too in different development in various sectors markedly at different regional levels which in turn helps for the regional economic growth and social development in the economy. "Space" and "Time" are the two elements to decide the Trade, Transport and Markets for accelerating the economic and social activity for which human resources are prerequisites. These three factors are identified as economic indicators for any economic development i.e. structure. process and stage in-turn increases the productivity, employment besides human development. Particular emphasis is being given to general transport structure of special network, consumer travel behavior and space preference. It will be particularly examined and identified as a feasible study in relation to the role of periodic markets, spatiofunctional interaction process in spatial organization, regional planning for regional economic and social development through the optimum utilization of natural and human resources.

Trade, Transport and Market systems are the three important factors which attracts essentially place (location) for interaction of trade, producers, consumers and their relations through the given transport network. 'Trade' is an economic activity identified by a day in a week and if it is continuous and stochastic in nature it is termed as "daily" market. This activity helps for earning a living and routine phenomena to people. This is purely an economic phenomenon over space and time for the growth and development of the economy as well as social development vertically and horizontally.

The 'resource base' is the base and basis for the manufacturing and retailing activity in a region. It is known that the economic activity assumes prime importance in the developing and developed countries and it is essential to identify in this micro level study helps to understand macro level changes in-turn that are required to acquire through infrastructural development. Recently, many scholars belonging to various social science discipline such as Economics, Sociology, Commerce, political science, public administration and Geography are concentrating and working on this theme. In western countries, this phenomenon was exclusively studied by many economists, sociologists. administrators, NGOs and geographers. This particular phenomenon was worked out by geographers in India and by Indian geographers abroad and the occupational structure of the people is mainly based on the existing resource based endowments of a place.

Rural Development and Urban Development are the two faces of a coin and "the economic planning" and "spatial planning" go hand in hand. To iron-out the spatial interregional variation, a viable and feasible plan should be prepared or augmented for implementation in order to achieve spatial, regional and vertical development either for a micro region or a macro region. The market infrastructural planning further leads to multi-level economic development in rural as well as urban space. This particular study helps to identify and under-lines the planning importance for the furtherance of the micro region's activity and the uncovered aspects will be covered through the studies such as general trade and transport, structure of transport networks, consumer's travel behavior and space preferences, role of periodic markets and rural development, urban development or urban shadow also initiate region the areas of spatio-functional interaction process in spatial organization and regional economic planning and also to suggest and implement the development plan for the economic development in a given economic environment depending on economic resources. It is understood that, "Transport is a nervous system and the inflow and outflow of consumers, producers, goods, services depend on good transport system which yield profit, interest, durability and utility for human being at the optimum level and it is compared to the blood circulation over space 'n' time "over a geographic space" (Quotation Mine please).

2. Literature

Indian and foreign scholar's contribution to this branch of study (as a backdrop) is in-exhaustive. Literature survey enables the researcher to find a gap in the study and it helps eventually to incorporate the some and the new light may be thrown to think about the problems as well as the process leads to unforeseen points to concentrate and find solution to the identified problem over space 'n' time.

Tanveer Ahmed, has presented the prospects of trade of Indian goods in Afghanistan and particularly highlighted the Indo-Afghan trade in the aspects (a) Indentified that Indian goods which have a prospect of expansion in Afghan market. (b) Availability of market on a competitive basis and (c) Evaluated the transportational problems. It helps to evaluate the civilization on the basis if the development of the transport artery which reflects on the overall economic development and cultural development of a region. Ravinder Pandey, has presented the transport chronology of Nainital District and considered it as a nervous system of a region and it makes a platform for the histogenesis and morphogenesis of this region. They have arrived at, "The impact of natural (physiographic) and cultural development is found pertinent in every period and the every section of the transportational development of the region". Chand. R. presented the economic development of India with the help of transport connectivity by using nodes and beta index and delineated the transportation zone. He has conducted the correlation method between transport connectivity and factors influencing it, shows that there is a perfect positive relationship between attributes of amenities ie. Population density, per capita income and transport connectivity. Thangevel and Anjana Vyas have proposed some basis; assumptions for a rural transportation system. They are (a) Central places are to t'- serve as the nodal points, (b) The rural roads are to be essentially access roads, (c) The locational utility has to be examined, (d) The roads have to be arranged in hierarchical order consistent with settlement hierarchy. Further opined that, "In developing the network structures the model takes into consideration according to the "Demo-Socio-Economic" characteristics of the region because the utilization and the future demand for roads depends on these factors besides this, it further considers the road pattern as an integral part of the settlement system.

The inflow and outflow patterns are important to note the movement of people and goods in which direction they are entering the market and where the goods are finally reach the destination. It further helps to understand characteristics of the hinterland of the region and its potentially over space and time. Umamaheswari has presented a study on the municipal vegetable market of Bombay city and in particular presented a comparative investigation of the three wholesale vegetable markets in Bombay (Maharashtra) in order to identify the localtional and morphological aspects of the above markets in relation to transport network.

R.K.Katiyar has presented some aspects transport connectivity and accessibility in a case study of Farrukhabad district, of Uttar Pradesh. This study examines and identifies that the 'accessibility' is a prerequisite for the unification of nodal areas with peripheries. The economic development of any area depends on multiple factors such as Physical, Economic, Social, Cultural and Political. For the development of these factors, 'transportation' is an essential tool, without which regional 'economic and social development' can't be feasible and plausible. This study has shown the impact on rural development with suggestions to correct the imbalances in transport sector so that it can serve the most notable objective of developing rural areas and provide minimum needs of the poor".

Singh B.B. and Mathur, A.K have presented a theme entitled, "Market surplus of commercial crops and its Directional flow in District Ghaziabad and the area is rapidly tending towards, commercialization in agriculture. Due to this increase in commercial crops in recent years and there is an apparent marketable surplus of the production of these crops is found in Ghaziabad and environs. Singh has identified the following indicators (1) Evaluated the marketed surplus of commercial crops, (2) Identified the surplus and deficit packets in the study area, (3) Detected the directional flow of the marketed surplus and its causes and (4) suggested a balanced reorganization of flow with requirement of the people".

Singh has postulated a transport network model of eastern Uttar Pradesh and found that the central places are focal points of the transport network, should have increased the localional utility of central places by providing easy access to village, lying in their hinterland. Road transport should be accorded highest priority in an agricultural economy because of its flexibility. Singh concludes that, "Rural roads should provide link from village to another as well as to local market and town and a survey on accessibility and develop a district level road work, which could be updated besides to develop link among the number of villages, maximum population and facility location".

Srivastava has examined the frequency of market place participants at market centers of the Tarai-region of Uttar Pradesh. The frequency of consumer's behavior takes into account in this context based on the models of economic theorists on one hand and in view of the individual perception of elements of time and distance on the other hand. The economic constraints overwhelmingly dominate other constraints "market Cycles" and "Market Circuits' are the two vital factors in any region, have a predominant role in the Socio-Economic pattern and in turn the development of geographic space for economic activity. Perishable and non-perishable goods have a persistent and pertinent (distance factor) difficulty to travel a long distance for disposal in a market. Mobility is an off-shoot to the perishable commodities and eventually faces the time censurability to travel over the space.

The market development of transport facilities, increased production of commodities, associated with dense population, clusters are followed by corresponding increase in the number of mobile vendors. In the given subset of market cycles, the relationship between the number of itinerant traders with the size of population involved in related market nodes seems to be greater than corresponding relationship of service areas of the same market nodes.

3. Role of Periodic Markets and Rural Development

Periodic markets have an important role in the development of rural areas. By introducing the socio-economic facilities and infra-development, rural areas can be developed besides the role of markets. The activities ranging from production to consumption of commodities within the specified framework of market, all diversified social behaviors and activities directly or indirectly link the entire system. Ii is the sum of aggregate of all dynamic actions and interactions at the crest of the social prestige of a region presents a unified case termed as markets.

The decimal variations present variations in the growth and development of the region. To illustrate the characteristics of the tributary areas served by market, three characteristics of market areas have been identitied by Srivastava, they are : (l) It has been found that market areas are arranged into five tiers of hierarchy which coincides with the hierarchy of market places, (2) The population, number of villages and the area served by the market places of higher orders is larger than those lower orders and (3) The market areas of the same order are comparatively larger in the western part of the region than the eastern parts. On the other hand the population and the number of villages served by the market places of similar hierarchical order are larger in the eastern part than the western part. These variations arise due to spatial characteristics of population, settlement and the nature of market places.

Anand has presented the spatial pattern of trading activities and the location of periodic markets, which follow a particular system of hierarchy. His study aimed at examining the location of periodic markets in fairly well developed agricultural taluk and analyzed the linkages of markets with periodic markets.

Hugar and Vijayarj have presented a temporal analysis of periodic markets in Dharwad district of Karnataka state and analyzed the market centres and market days and compared the different periodic market system in different parts of the world. They have investigated (1) Spatiotempora l spacing of periodic market, (2) The extent of optimally of space-time sequence to people who use them, such as consumers and traders, (3) The space-time sequence analysis has been applied to markets in order to test the each day of the market week for 1931-81, and the consumer and traders hypothesis in integrated days and non-integrated days of the market week.

Vishwanath presented the distribution pattern and classified the market centres in Awad of plain (North) by using the NNA) on the basis of periodicity, the market centres have been classified. He further opined that "According to the means of transportation, the status of the market centres and the nature of agricultural produce exchanged by them".

Pandey and Siwarli have presented the implication of topologic measures of transport network in the integrated area development. In a case study they have observed that the elementary topologic measures of the transport network which include Beta. Gama, and Alpha indices cyclomatic numbers and average parth length have been applied and computed for Tarai and Bhabar region. He opined that, "Measurement of association between transport connectivity and central variables like population size, educational, medical and postal facilities suggest accepting the hypothesis which were found significant at 99 percent composite index (C.I) and many be used to plan for integrated area development in Tarai and Bhabur region.

Keeping in view of the earlier studies, now the prime concern of the investigator is to conduct an intensive survey over an extensive geographic area by identifying the gaps which are uncovered in some respects will be covered in this research studies in this case study of a district from Andhra Pradesh may be taken-up in order to identify and recognize the variables which present the variations in the levels of spatial development are important aspects of this study. The aspects such as homogeneity and heterogeneity in particular will be identified eventually in this study will find a place and solutions are offered to achieve 'integrated' development of this region by linking the road network with potential market towns. However, the economics of market centres with a good transport network can function as good service center, from where the rays of development trickle down. A good programme will be progressed to develop market centers to strengthen the periphery areas faraway places with the market centers tightly will create a 'fusion' and infuse economic activity over space 'n' time to achieve 'integrated' development.

4. Objectives of the project

1. To study the existing market facilities and examine the pattern of hierarchy among the market centres.

2. To study the trend pattern of population at the micro level (i.e) Mandal level (2001 to 2011) and estimate the future population growth for 2021.
3. To study the resource appraisal in the fields of agriculture, industrial potentialities and infrastructural patterns.
4. To rank the central places according to the centrality score of the mandal/settlements and composite index will be constructed.

   Besides the above, the investigator intends to present the strategy for the development of urban centers/market centers for the optimum utilization of their' services for the development of the district as a whole and also to find ratio between the urban and rural market centers/settlements of the region by reducing spacing pattern and in turn increase the potential population to interact with the proposed market centers for achieving topologic optimality by suggesting new market centers.

5. Hypothesis

It is hypothesized that the study area is uniformly developed (i.e) homogeneity is being maintained in the development process. If not so, what are the heterogeneous factors contributing to the variations in the levels of development and also to iron-out such contributory factors for attaining, obtaining homogeneity and wipeout completely the heterogeneity in the study area. If we quantify the infrastructure facilities for socioeconomic development, the quality of life of people may be stepped up, improved or alleviated to a large extent over a period of time.

6. Research Design

Nalgonda district have 59 mandals covering an area of 14,240 km2 with the population of 28,52,092 (1991). There are 27 factors relating to physical, Socio-economic and infrastructure factors which contributes for the overall development of Nalgonda district. This is to say, that (27×27 this is further factorized each 9 factors form into a subgroup which causes physical variation, socio-economic variation and infrastructure variations [i.e. 3×9=27 (9+9+9=27)] it will be assessed that both vertically and horizontally, the variations are to be identified even within the subgroup. That is 9/9=1 or less than l. If 3/3=less than 1, and then we assess that particular subgroup has its roots, causing variation to development and if it is above I, to some extent the sub-group is developed but not fully developed.

In the same way, we have to examine that 27/27=1 or less thariwithin the given framework of indicators and development.

7. Methodology

The basic theory to be used in this study is Christiller's Central Place theory at the primary level including the Administrative Principle K=3 and Market Principle K=7. At the secondary stage, the factoral study will be conducted in order to identify agricultural patterns, market patterns, the consumer's travel pattern and human behavior to act, react and interact with the market places of the space preference so as to satisfy the needs of the consumer.

8. Tools for collection of data

The structural questionnaire will be designed specifically for the data collection, the primary level. The relevant data will also be collected from all the mandals in the study area. It is to be obtained through interview with local people/Village Development Officer (VDO) and the secondary data will be collected and supplemented from various government agencies to facilitate the researcher/investigator to improve them in order to identify the variables for further analysis.

Plan Period

The plan period proposed in this study covers a period over ten years (i.e.) from 2010-2020. Twenty per cent of the mandals will be chosen for this particular study in border to identify the pattern of development as well as market centers, which enables the investigator to analyze the data.

Significance of the Study

The survey, analysis and results would definitely be useful to the district machinery to enable them to implement some of the recommendations / suggestions which may be incorporated in planning and development of this district. It is evident to state that

Nalgonda district has been selected for study since 1956 no geographer /economists relating to transportation and market as principal resources to harness the final products to the market centers, to identify the problems and (search) find solution to the existing problems as well as future problems. Over a period of 5 census years spread over 50 years, have passed and the roots of substantial land sustainable development process has not been augmented. It is highly desirable and significant to conduct an in depth survey over an extensive geographic area covering the uncovered aspects such as, Flood prone / Drought prone / Cyclonic prone / River prone and Non River prone, irrigated and un-irrigated areas and their inherent characteristics of land-uses, occupational structure linked with the economy of the people besides amenities available (less) for them in the total framework of the economy and the livelihood and likelihood of market centers to be created to find a place for the agricultural products (Agri-Marketing) in the irrigated and un-irrigated landscapes.

The study area is considered to be a river prone and non-river prone areas besides climatologically, edaphically, pedagogically, economically and culturally the primary economic activities/ secondary activities and tertiary activities are deep rooted causing variations. The cumulative effects of high/low rain-fall, deficiency in nutrien\ soil for carrying out agricultural operations and inadequate irrigation facilities with a resultant low yield makes agriculture a chance in un-irrigated land-holdings leads to a distress to the people and the economy. Hence to cope up with the increasing development strategy on agriculture side, there is need for planned development of the market centers (Agricultural market centers) also with the provision of more infrastructural facilities, for economic development. Hence there is a need to take up such studies to augment the need based planning to incorporate in the district planning (Micro planning) unit level for achieving economic and social development.

10 Relevance to the present day problems and needs of the society and the Country

Mandals and market centers in the proposed district study are to be analyzed III relation to its service, administrative functions and marketing principle of Christallers Central place Theory and also to find an alternative methodology that is factor analysis technique to supplement the existing theory and its operation in the study area. All the urban service centers and market centers are to be arranged and assessed in order to ascertain and establish the hierarchy. All the settlements should be arranged into an ascending order of its population to enable the researcher to apply the concepts of Central Place Theory.

The urgency warrants the researcher/investigator to conduct a survey in order to identify, analyze and interpret the environmental hazards of the study area and also to find possible and plausible solutions to overcome the situation. It is the primary duty of the individual to identify the most important and urgent needs of the society belonging to the environmental/ health hazardous situations and also to create a network of Physio-Economic and institutional infrastructural development over period of time. (Spatiotemporal aspects).

11. Likely Contribution to Knowledge

This knowledge contributes at large for solving some of the specific needs of the society and the country through this planning. If we consider the district as a micro level unit, the meso unit is the state and the macro unit is the nation. Though this work appears to be applied in nature, the basis of planning at gross roots level is much felt by the society. It is important to note that the regional needs and aspirations of the people should be enlisted/chartered and priorities should also be identified at the micro level. Ultimately, the regional needs should be incorporated and reflected according to the priorities of the society and find viable solutions to identify the problem is a must. Regional economic and social development could be achieved, if the regional needs are identified to this survey and the result-oriented solutions are to be obtained over a period of time for application. It is noted that through transparency, efficiency could be obtained sufficiently, if the people participate in the planning process for achieving the desired and targeted development, as Government either state/centre has been trying to involve and induct people in planning and enable to contribute their share (If people's efficiency is matched with the government' s economic share) for achieving the overall and integrated development as well as special stress on Sectoral Development.

Finally, this will help in-turn ultimately to reach the spatial economic and social development through transport-lines and market centers. It is necessary for people to act, react and interact with government's policies to achieve spatial-economic, social and cultural development through this micro level planning machinery at the district level, keeping in view the principles of planning at the optimum level to achieve the overall development of the economy and the society at large over a period of time is required.

Reference

Dr. Narasimha Murthy K.L., Research in Geography: A Survey (pp. 73-86), published by Ashish Publishing House, 8/81, Punjabi Bagh, New Delhi-110026.

Part - C
Geography

SKUCET Home (index-html) / Departments / Geography

About Department
Head of the Department
Faculty
Courses offered Activities

About Department

Year of Establishment: 1985

Brief History of the Department:

The Department of Geography was started in February 1985 in Sri Krishnadevaraya University with full financial aid by UGC with a view help the study of available natural resources. Biotic resources, Physical resources and Mineral resources of any region in India and with special reference to Andhra Pradesh and Rayalaseema Region. The students of M.Sc., M.Phil. and Ph.D will be imparted knowledge of the resources, cartographic mapping and interpretation of topographic and remote sensing data. Geographers would be useful to society to serve in Micro, Marco and regional planning, land-use studies and in environmental programs.Of late, the students of Geography are carrying out resource evaluation studies using GIS techniques for micro watershed management, mandal level planning, district level planning, regional planning, state level planning and national level planning.

Contact Us:

Chairman BOS: 255851 Head: 255755 Office: 255756

Head of the Department:
Prof. A. Krishna Kumari
BOS Chairman:
Prof. A.Krishna Kumari

Table 81

Faculty

Name	*Designation*	*Mobile*	*Office*	*E-mail*
Dr. M. Sambasiva Rao	Professor	9949986444	248048	Msambasivarao2006@gmail.com
Dr. A.Krishna Kumari	Professor	9949774986	255851(0)25531	Krishna a31@yahoo.co.in

Courses Offered

Courses offered by the Department: M.Sc, M.Phil, Ph.D

Areas of Specialization:

Geomorphology, Hydrology, Agro-climatology, Agricultural Geography, Ecology and Environmental studies and Medical Geography.

Activities

University Colleges

SKU College of Arts (skucoa/index.html)

SKU College of Science (skucoa/index.html)

SKU College of Engineering (skucoa/index.ac.in/)

SKU College of Pharmacy (skucoa/index.html)

SKU College of Education

Useful Links

Centre for Skill Development

Entrepreneurship & Incubation (http://skuskills-org.webnode.com/)

Alumni &Friends (alumni.html)

Campus Map (skumap.html)

Contact SKU

Table 803

Sri Krishna Devaraya University
Ananthapuramu-515 003 (A.P.), India
List of Ph.D's Awardees Department wise (2010 to 2016 April)

Sl. No.	*Name of the Candidate*	*Name of the Research Supervisor*	*Name of the Department*	*Date of viva- Vove*
373	K. Kusuma Kumari	Prof. M. Sambasiva Rao	Geography	04-07-2015
374	YJayalalitha	Prof. A. Krishna Kumari	Geography	30-07-2015
375	Yannam Chinnaiah	Prof. A. Krishna Kumari	Geography	09-09-2015
360	T. Somasekhar	Reddy Prof. M.Sambasiva Rao	Geography	06-06..20 II
361	Karunakara	Rao M Prof. A. Krishna Kumari	Geography	24-02-2013
362	M. Raveendra	Prof. M. Sambasiva Rao	Geography	23-06-2013
363	K. Somanna	Prof. M. Sarnbasiva Rao	Geography	19-08-2013
364	Anjaneyulu B	Prof. Y.V. Ramanaiah	Geography	13-11·2013

In Chapter-8 Part-C: S.K. University - A Case Study from Andhra Pradesh Contribution by the number of Individulals, Department of Geography, Srikrishna Devaraya University, Anantapuram, Est. 1985.

Table 8.2

Sl.No.	*Name of the supervisor*	*Awarded Degree (Number)*	*Ph.D Projects sanctioned & Completed*
1.	Prof. E. Swaminathan	03	
2.	Prof. M. Sambasiva Rao	24	
3.	Prof. Y.V. Ramanayya	12	
4.	Prof. A.Krishna Kumari	07	
5.	Dr. K. Ravinder Reddy	04	
	Total	50	

Analysis

It is seen from table-Svl, Department of Geography, S.K. University, Anantapur has established in the year, 1985 and has brought out 32 batches of M.Sc (Geography) students and the major specializations are geomorphology and agricultural Geography of the M.Sc (Final) year and the staffing pattern has also been stated So, far 50 Ph.D degrees are awarded to the candidates besides M.Phils. Major research project's were also sanctioned to the staff members and have contributed to Geographical knowledge since 1985 to date.

Conclusion

It is concluded that many variations are observed in this study and analysis of paper contribution and the author has presented the Bibliometric Analysis of the research papers published in journals - an evolution and evaluation in chapter-Z, based on 5 year.s period (2013 - 2017) and Bibliometric Analysis relating to micro level survey has been presented in Chatper-8 and finally, summary and conclusion are presented in Chapter-9 in order to identify the set patterns of research paper. It is suggested that A Geographical Journal should be started at least at a University Centre i.e. Department of Geography, Andhra University Vishakhapatnam Department of Geography S.V. University or Department of Geography, S.K. University, Tnanta p. in Andhra Pradesh state in order to spread the roots geography and development in the State.

Notes

1. Bhat. L.S.. *"Micro Level Planning - A Case study of Karnal. Haryana"*. K.B.. Publications, Delhi. 1976.
2. Prakasa Rao, VLS, *"Development strategy for an agricultural region - A case study of Muzaffarnagar District. UP"*. Concept Publishing House, New Delhi. 1977.
3. Surender Singh & Chauhan. V.S.. *"Regionalization for Rural Development and Planning"*. Published Independent Publications Pvt. Murat. 1982, pp. 3-4.
4. Reddy, Y.V. and Kosal Ram, *"Multilevel Planning in India, Processes and Perceptions - A Study of Planning in Andhra Pradesh"*. Published by Centre for Economic and Social Studies. Hyderabad Book Links Corporation. Hyderabad, 1988.
5. Mahammad Afzal. *"Spatial Framework for Development, Administration and planning"*, in Multi level planning in India, Processes and perceptions. Published by CESS. Hyderabad and Book links Corporation. Hyderabad. 1988. pp. 545-559.
6. Hassan. Mohd Izhar. *"Regional Inequality in Orissa: Some Emerging Issues,* pp. 10-19, *Trends.*
7. Nayak, L.T., *"Regional Disparities in levels of Development in Bellary District - Karnataka"*. pp. 37-47, Indian Journal of Regional Science, 2005, No. 1.
8. Diwan. G.R. *"Urban & Regional Planning Activities in Maharashtra"*. Indian Journal of regional Science. No. 1. 2009.
9. Ahmed, Naushad, *"Regional Imbalances afford security in Northern Bihar Plain"*. Indian Journal of Regional Science. 2007, No.1. pp. 50-56.
10. Khsn, Nizamuddin, *"Spatial Patterns of Agricultural Development in Murshirabad district, West Bengal"*, Indian Journal of Regional Science, No. L 2009, pp. 72-78.
11. Ali. Mohd Julfikar. *"Micro Regional Disparities - Analysis ofAgricultural Crop Productivity"*, Indian Journal of Regional Science. 2009.
12. Power. 0 .11. *"Geographical Analysis of Pressure on Land in Maharashtra"*. Indian Journal of Regional Science. No. 1. 2009.
13. Power, C.T.. *"Role of Market Centres in Regional Development: A Micro Level Analysis"*. Indian Journal of Regional Society, 2004, No. I, pp. 103-112.
14. Imam, Shahid, *"Economic Impact of Indira Gandhi Canal Projects Western Rajasthan"*. India, Indian Journal of Regional Science. No. 1. 2006. pp. 45-54.
15. Power. C.T. *"Regional Disparities in Levels of Human Resource Development in Southern Plateau Region of Maharashtra"*. Indian Journal of Regional Science, 2009. No. 1, p. 92.
16. Pal Manas, *"Response of Marketed Surplus: A Case Study Qf Paddy in Burdwan District"*. Indian Journal of Regional Science, 2009. No.1.
17. Das. N.. *"Hierarchy of Market Centres in the Rural Hinterland of Cuttack. Bhubaneswar. Orissa"*. India, Indian Journal of Regional Science. 2005. No.1 pp. 120-121.

18. Jay Sree, D & et.al., *"Disparities into Levels of Development in the Districts of Gujarat".*
19. Ahsis Sarkar & Das, Sandip, *"Agricultural land Resources of Hoogly District".* West Bengal.
20. Shukla, S.C. & Dhagat. S. *"Land Utilization in Madhya Pradesh: A Factor Analysis Approach",* Journal of Rural Development, Rajendranagar. Hyderabad. 2006.
21. Rao, T.N.A., *"Balanced Regional Development: A Development Plan for Karnataka State. India".* Published by Academy of Regional Science and Humanities, (ACARSH). Belgaum, India, Veena Rao. 1993 pp. 4-5.

Chapter - 9

Summary and Conclusions

'International Geographical Research - A Survey' is an important volume which has been identified and useful for young researchers / students and teachers to know the current trends. Traditional branches in Geography are as follows. They are (1) Physical Geography / Principles of physical Geography, (2) Economic Geography, (3) Urban Geography, (4) Rural and Regional Geography, (5) Geographic thought. The withered branches of Geography are as follows: (1) Social Geography, (2) Political Geography/Geopolitics and (3) Cultural Geography. They are replaced partly by R.S. GIS quite recently.

National-body of geographers have identified some more branches besides the above. They are: (1) Human Geography, (2) Geography of Resources, (3) Gender Geography and (4) Medical Geography. Besides the above branches some more branches are identified but not much has been done. They are (1) Plant Geography, (2) Zoo-Geography. But, in the international field most popular and populous branches are: (1) Physical Geography, (2) Economic Geography, (3) Urban Geography and (4) Rural and Regional Geography.

The author has made an attempt to present the above four most popular branches for review purpose in order to identify and recognize the gaps of research and suggested the 'areas' of research that are to be taken up by young geographers for pursuing M.Phil/Ph.D Courses of study.

In the first chapter, the author has presented the subject geography and its relation to physical, economic, social, cultural and regional aspects of the existing current phenomena upon the earth and pleads that it would be visible well in the form of the action, reaction and interaction by men within a special and identified environment. The nature has been nurtured the physical environment into three stages to that of a river cycle (i.e.) young, mature and old stage and have been identified into that Pattern which relates to structure, process and stage.

2. Physical geography: A Survey (International Level) deals with quantitative Geomorphology and Climatology in Chapter-2 is presented.

1. Terrain studies
2. Paleoclimatilogical simulation and climate change
3. Remote sensing
4. Land.use and land-cover dynamics or land-use geography
5. Gaps in literature; and
6. Suggestions for future development

3. In the third chapter, the entire text of the survey has been designed and organized with the following sub-heads. They are:

1. Economic - Geographic Situation
2. Ecology
3. Vegetation
4. Livestock
5. Human Environment - interaction and sustainable development
6. Land-use change
7. Economy
8. Remote Sensing
9. Geopolitics
10. Gaps in literature
11. Future areas.

In chapter -4 the author has attempted to present 'Urban Geography: A Survey' (International Geographic Research the current scenario of this branch and its elaborative content is as follows. They are:

1. Urban Ecology and environment
2. Urbanization
3. Methodology - Globalization & Technology
4. Urban Growth - Urban Sprawl
5. Urban Poverty

6. Immigration & Migration
7. GIS methodology
8. Culture
9. Gaps in literature and
10. Future areas of research

5. The author has presented the themes such as the rural Geography, Regional Planning and Development Planning/various aspects of rural Geography, Regional Planning - and Development Planning research studies with the following themes in a total framework and format. They are:

1. Cultural elements
2. Population
3. Methodology
4. Migration
5. Millennium Development
6. Cluster Development/Regional Innovation
7. Gaps in Research and
8. Future area; studies

The Bibliomatric analysis of papers (international) has been analysed in the 6th chapter. The Chapter deals with Indian's contribution to Indian geographic research including Bibliometric Analysis and the themes are as follows. They are:

1. Environment & resources
2. Physical Geography
3. Economic Geography
4. Urban Geography

5. Rural Regional Geography and Rural Development is presented in this chapter from second to fifth, core chapters of first series are presented relating to international geographical research exclusively from 1-85 numbers (Eighty five papers), where as in the second series, themes are presented relating to Indian's contributions to Indian geography planning from nation to state and from state to district/mandal/village level studies and also Indian's participation in the 33rd international geographical congress, Beijing, China, Asia, 21-25, August 2016 besides the papers from Indian Geography Journals (i.e) in particular, referred to in the text are:

1. The Geographer Aligarh
2. Transactions, Pune

3. Annals of the National Association of Geographers, India, New Delhi.
4. Indian Journal of Regional Science (IJ.RS), Kolkotta, West Bengal, Thematically spans the papers related to state levels, District study and Taluk study representing micro level units in the hierarchy of international (Macro), National (Meso) and state and sub-state belongs to smaller units in the specified form and they are as follows:
    1. Environment & Development
    2. Geographical Landscape
    3. Geomorphology
    4. Geospatial technologies
    5. Remote sensing and GIS
    6. Urban Geography - Impact of Human Interaction on Geo-environment (District study)
    7. Urbanization
    8. Urban Slums - Macro study - Bhopal city
    9. Comparative study of slums
    10. Rural Urban fringe
    11. Land-use and Agriculture (District and State Study)
    12. Kashmir Valley
    13. Crop Diversification (State level Study)
    14. Agricultural Development (Taluk Study)
    15. Garden Culture - Tea (North-East Region of India)
    16. Sustainable Agriculture - Haryana (State Level Study)
    17. Agriculture in Drought Prone Region (Hamirpur District, Himachal Pradesh)
    18. Healthcare System and Rural and Regional Development and also presented the bibliometric study at the National level.

Chapter 7 deals with bibliometric analysis of the research papers published in Indian Journals (as referred to 21 papers), and

Chatper-8, deals with micro level planning and integrated area development - A Model study and also presented the bibliometric methodology@ micro level units.

Chapter-9 deals with summary and conclusions. Besides presenting a special research paper, "Resource base and pattern of Development: A case study of Vegeswara puram, Talla Pudi Mandal of West Godavary District, Andhra Pradesh, India.

Conclusion

1. It is concluded that in the subfield of physical geography 'Geomor-phology, there is every need to work on the river basin morophormetry with climatological parameters as it will be evident that climate play an important role in sharpening and shapening the landforms and topography, as it makes the geomorophologists to further specialize on climatic geomorphology likewise. Soil and vegetation should be clubbed with temperature and precipitation variables in urban to identify recognize and understand the inter relationships for growth and development.
2. Further concluded that urban climatology especially Micro Climates which have a bearing upon urban ecology and urban society. Heat Islands are not the recent phenomena, but existed in previous years also, but the violent structural composition has a silent impact with its unforeseen speed and ultimately pattern has telling effect upon people residing in urban and urban fringe areas.
3. It is noted that air movement both vertical and horizontal have its impact on the lives of human being and its impact on humans and in a way leads to diseases) Particularly respiratory and lung diseases that environment supports for its existence.
4. Air pollution, water pollution and soil pollution are quite important and its impact on human society over a period of time is important for undertaking a research problem. Besides this, sound pollution should be considered as one along with other, creates problems in the name of diseases to human living in different settlements. Studies should also be conducted in this branch which enables the planners to tune/formulate policies and implement them for urban welfare, urban development for people.
5. Socio-Economic and cultural elements which contributes to the spatial patterns of settlement, urban densities and urban infrastructure and its deficiency in the quality of life amongst the urbanites specific planning to the urban environment and plans should have to be designed and implemented without deviations.

Social overheads in the form of urban infrastructure should be provided which results not to face the urban congestion due to rural - urban migration. As a result the urban environment will be changed to a large extent.

6. 'Micro Urban Plan should be designed to combat the deficiencies in urban human life. Urban poverty is a phenomenon which leads to dangerous situation and eventually, it creates unemployment and social disorganization due to rural-urban migration. Major municipal Corporations have to take necessary steps to eradicate this phenomena and find solution to this urban problem. This is an urban challenge to planners in India.
7. Integrated Area Development plans should have to be designed and planned so as to create 'social touch' among the urbanites and urban employment schemes should be implemented in order to achieve improvement process and development in multiple fields.

In the Chapter the author has identified the authorship pattern of the international reputed journals and this pattern should have to be redesigned and welcome new feature among the researchers that individual contribution is more essential, warranted and multiple authorship of research papers should be discouraged. As the growing technology, pin points the researchers to follow the current trends in geographic research especially in the international standard, publication. Due to the application of RS & GIS, many research papers are published with multiple authorship than individual contribution and authorship of a good research paper which could be seen in the physical geography and urban geography. In the other branches of geography, single authored papers outweigh the entire Scenario both at intentional and National level.

Finally, the author makes a special and sincere request that the researchers should have to aim high and work high for the ongoing progress of research scientifically with the techniques of their own to develop the theme and find genuine solution tò the selected problem by the researcher. The governance may be able to take up the fruits of research, in one way, academic contribution may help the governance to visualize and find suitability and applicability of these particular methods to a growing problem to find solution amicably. 'Social Science' research problems should be selected for study and find scientific solutions to the urban sprawl as well as urban poverty. To eradicate urban poverty, scholars should have to plan, design the area specific and population specific plans through this spatial (area) integration and functional integration are the two aspects which are vulnerable, (valuable and feasible the governance may be able to implement the plan in toto for the welfare of the urban society. Research should seek the welfare of the society, thus urban planning should have to be tuned in tune with the needs of the urban society and fulfill the aspiration of the urban voter. A research papers have been presented, at the end, for the beginners of M.Phil and Ph.D. Relating to resource base and pattern of development A case study of Talle Pudi Mandal of west – Godavary District, Andhra Pradesh, India.

Prithivi Ratna, Param Bhushan Prof (Dr) B. B. Shukla,
M.A.; M.L.Sc. (B.H.U); Ph.D., D.Litt (LIS-1st in the World), LOTW.
Professor of Information Technology & Environment (HC)'
Founder Editor, IJILtS & PBSAMA;

M : 9439282882
At: Nua Sahi, Po : Nayabazar,
Cuttack - 753004 (Odisha), India

Ref. No 81.....,   Date : 1.10.17

Dear Dr. Murty ji, Good morning.

I am in receipt of your letter and the article titled "Current Trends in Geographical Research – Applied Bibliometric Methodology" for publication in IJILIS. I have gone through the text of your article and found it suitable for our readers. Accordingly I am publishing this article of yours in V30, N3-4, 2017 of IJILIS and soon after its publication, I shall send you a copy for your reference and use.

Once again I thank you very much for your interest in our journal "IJILIS".

Yours Sincerely,
B.B.Shukla
PRITHIVI RATNA PROF(DR) B.B.SHUKLA
Ph.D,D.Litt (LIS-1st. in the World), LOTW
Founder Editor, IJILIS,
At-Nua Sahi, P.O-Nayabazar
Cuttack-753004 (Odisha) India

To
Dr. K. L. Narasimha Murty,
305, Manikanta Sai Sudha Apts,
Meena Cafe Lane,
Narasiguda,
Hyderabad- 500061

Prithivi Ratna, Param Shushan Prof. (Dr) Sana Bihari Shukla, The Highest Qualified Information SCientist of the World during the 20th & 21st Centuries; the President, Universal Service Society for Human Life and Environment (HC); the Selfless Social Activist, Internalionallnformation Scientist, Educationist, Environmentalist; Humanist and the Protector of Fundamental Rights of the Citizens, particularly of the poor & downtrodden people and innocent other Creatures of the World.

RESEARCH PAPER ONE

Bibliometric Analysis of the Geographical Journals: 2011-2017

K. L. Narasimha Murthy*

1. Abstract
2. Introduction
3. Literature Review and Gaps in Literature
4. Objectives
5. Material & Methods
6. Analysis
7. Typology of Articles
8. Ranking of papers
9. Findings
10. Conclusion

1. Abstract

An attempt is made in this paper to identify and recognize current trends in geographical research through applied bibliometric methodology. The traditional branches of geography are: (1) Physical Geography, (2) Economic Geography, (3) Population Geography, (4) Urban Geography, (5) Social Geography and (6) Regional Geography.. Indian Geographical Research has been indentified and recognized through the parameters, such as, (1) Single Authored papers, (2) Two authored papers, (3) ;Three authored papers and (4) Multiple authored papers. The published papers were classified over a period of five years. At the second stage all the authored papers were given / assigned weightages, such as (1) 100 Scored points, (2) 50 Scored points, (3) 25 Scored points and (4) 12 scored points. All the six branches papers were to a tune of 162 and have arrived at cumulative index (i.e.) weightage X total number of papers for each branch has been calculated.

'Ranking method' has been applied in order to identify & recognize the role of technology such as remote Sensing (R.S) and Geographic Information System (GIS) methods and the remaining papers were classified as traditional typology (T.T) Papers both qualitatively and quantitatively. At the end, the author has suggested some valid measures to encourage fresh research scholars to the latest methods of research and analysis in geography. The study is limited to geography journals in this paper.

Key words: (1) Geo- environment, (2) Land-use and Land-cover, (3) Resource Evaluation, (4) Peri.Urban Region, and (5) Geospatial Technology.

2. Introduction

Eratosthenes coined word 'geography' from two words. 'Geo' means 'the earth' and 'graphien' means write or describe. The description about space, Eratosthenes.

195 B.C prepared good maps, which are said to be known as the earliest maps. The dictionary meaning of geography as the science that describes the surface of the earth and its associated physical, biological, economic, political and demographic characteristics entails a vast scope of learning, teaching and research in geography. It is to state that 'geography' is the study of the assemblage of physical, economic, social, cultural and regional level aspects of the existing phenomenon and environment (quotation mine please). The entire research activity takes place into three stages. They are I) Structure, 2) Process and 3) Stage.

"The study of quantitative assessment of published papers and their citations are coined as bibliometric study. Bibliometrics is the study of journal citations which in turn identifies the relationship among the cited articles. Their association brings out the logical, perhaps the link with the studies of past, present leading to further problems and themes of research. In this process it also makes quantitative analysis of published Iiterature"[1].

In the present paper, the author has made an attempt to the study of qualitative assessment of published papers and their citations relating to geographical research. A total number of 162 papers are classified and analysed. They are: (1) Singe authored papers, (2) Double authored papers, (3) Three authored papers and (4) Multiple authored papers. They are related to major branches of Geography. They are: (1) Physical Geography, (2) Economic Geography, (3) Population Geography, (4) Urban Geography, (5) Social Geography and (6) Regional Geography. At the second stage, the papers were given weights such as, 100, 50, 25 and 12.5 point score. Total number of single authored papers were multiplied by the respective given weight to get the total scores. All the scores were classified and were given ranking to arrive at an index. Such indices help to identify the status and structure of the capacity, capability and caliber of the author to identify the indepth of knowledge. Sub-fields wise distribution of papers are also presented. At the end, some suitable suggestions are listed for young researcher's research activity and personnel to carry-out their research activity successfully.

Bibliometric analysis of the journals, during the period of survey (2011-2017), focused out, that out of 162 papers, 71 articles belong to single authored articles: while the rest were contributed by the joint authors. It indicates that most of the Indian Papers and almost all papers covered by Indian authors on the Indian landscape were taken up for review. This bibliographic study of geography journals, wherein, the collaboration and geographical distributions were at length discussed and suggested the future agenda of research for young and energetic research scholars of age group of 25-35 years.

2. Literature Review and Gaps in Literature

1. Role of Technology in Physical Geographical Research.

Land-use and land-cover study is an important aspect in understanding the interaction of human activities with the geo environment. In this paper, an attempt has been made to study the changes in land-use and land-cover transformation s of Vishav watershed over 20 years period (1990-210) by using remote sensing and GIS which play an indispensible role in detecting, analyzing, monitoring, recognizing and comparing the land use / land cover changes and trends.

2. Role of Technology in Economic Geographical Research

"In ground water resource evaluation, the quality of ground water is of nearly equal importance to quantity. The study of ground-water quality involves a description of the occurrence of the various constituents in ground water and the relation of these constituents to water use and also to notice that quality of groundwater varies with respect to time and space. Further examination of water in the field to determine its approximate chemical character is often useful. Having the water tested by a reputed laboratory is the first step in this process. It deals with the quality of groundwater for irrigation of a part of the Ganga Basin, Varnasi district. He has assessed that Ram Nagar shows a very high electrical conductivity (>30000 min/mls/cm) and is of poor quality under C.S. Group which is unit for irrigation. These require proper salinity treatment before being used in the fields. Ground-water in the rest of the part of the Ganga Basins of Varanasi district is moderate to good in quality[2].

3. Role of Technology in population Geography

The authors have presented that, "Cities in India ranging from fast growing Mega Cities to Small Towns are facing the dual pressure of rapid population and inadequate infrastructure. The multiple dimensions of sustainability- economic environmental and social aspects are inter-related. Sustainability of slums have been expressed transformation of economic growth into social and environmental achievements (especially in aspects related to availability of water, toilet, sanitation facilities, electricity, health and environment and Basanth Colony a slum in Bidhan Nagar, a satellite town of Kolkata as the case study in this regard. The empirical study is based on questionnaire survey among the slum dwellers as well as secondary data analysis. The objectives of this study are to find out-environmental and social changes being faced by these slum dwellers. Also it identifies the short-coming in the provision of basic services of living and finally suggests ways to reforms towards sustainable provision within the administrative and institutional framework[3].

4. Role of Technology in Urban Geography

The authors have provided this study and it proposes to assess the presence of providers and the preparations of beneficiaries in terms of obtaining degrees and diplomas from the even increasing numbers of such institutions. This study is mainly based on Kolkata Peri-urban region. They have concerned that the stupendous rise in the number of institutions does not imply an equivalent increase in quality of education and professional knowledge. "It is a matter of great concern that poor families pay heavily for services that don't necessarily ascertain employment or even a considerable increase in the value of education. It is therefore, proposed that all these institutions be assessed and monitored to the private institutions may not be allowed to take their student accessibility for granted and that they continue their business only if they satisfy certain criteria. "The authors have collected data and have been qualitatively analysed and certain parameters mapped with the help of GIS Software".[4]

5. Role of Technology in Social Geography

S.P. Chetterjee opined that, "in the last quarter of the century digital revolution impacted mapping practices, soon came series of geospatial revolutions which radically transformed the whole practice of mapping and map...use. Now, ISRO and its various agencies command this technological development and is endeavoring to put it practical problem solving it the societal plane. NSDI is net-working and co-coordinating all major G.I. supply agencies so that all G.I. data wide transact through the India Geospatial[5].

6. Role of Technology in Regional Geography, Regional Planning and Development Planning Studies.

V. Rangan and G.R. Parthasarathy, have presented in their task work "Identification of Service Facilities of Urban Centres in Thene district, Tamil Nadu. A geospatial Analysis", and looked into the issues of service facilities of urban centres in Thane district in the context of regional development from geospatial perspectives. It is to be noted that it is a study of small and medium towns. Small and medium towns help the population to easily access the needed services, thereby reducing the pressure on the larger towns. Thene district is an inland area in *Sin* Tamil Nadu, and towns have belonged to the categories of class-2 to class-5, they normally provide functions to market, health, education and recreation. Methodology of simple statistics is followed for the required analysis in the study. Mapping and spatial analysis for the present exercise is carried out with the help of Arc GIS 9.3 and ERDAS Image 8.7[6].

Bibliometric Methodology - A Review

Rarnesh, LSRCV and Naga Raju, AVSS have presented a paper on "Publication pattern in Indian Journal of Information, Library and Society, 2007-2014. A Bibliometric study, deals with the discussion of the publication pattern of the most prestigious Indian journal of information, library and society over a period of time. This article includes feature-wise analysis of the journal. The study on authors emphasizes the worthy encouragement given to new authors with systematic analysis. Overall, it indicates an organized team work, keeping the National Journal in the fore-front of Indian Journals with dynamic stewardship for the past 27 years"[7].

M. Padmaja has presented a paper on "Bibilometric Analysis of the Journal of Rural Development: 2009-2013" deals with an evaluation regarding number of papers; citations therein appeared in the publication of the quarterly Journal of Rural Development. She has focused on 20 issues of 5 volumes during the period 2009-2013. She has considered the authorship pattern, estimation of citations and volume-wise spread of papers, strength of citations and subject-wise analysis of published papers. The bibliometric study of the journal is analysed and findings are presented with the help of suitable statistical techniques. The study is limited to the journal of Rural Development, the Prime Publication of the National Institute of Rural Development[8].

Simhachalam, J & Narasimha Murty, K.L. have presented an Evaluation of Research methods in Library and Information Studies - Bibliometric study: A Survey (2005-2009). They have attempted intensively over an extensive geographic area in this paper to evaluate the evolution process of structure, process and stage elements in library and information studies to identify the above processes, the data base is created and classified according to its structural components and its influence and also conducted a survey for a period of 5 years relating to bibliometric studies, evaluated from the journal, "Annals of Library and Information Studies". Aggregation and Disaggregation of Knowledge that has been Spread over the space from micro level to the macro lev.el and also to disaggregate and disseminate the spread knowledge which has been presented to the students, teachers, research scholars and also for the seasoned officers of Library and information studies to know the current study in LIS. Bibliometric study has been carried out. Out of 140 published articles, 20 papers were thematically presented relating to Bibliometric study. At the end, the potential areas of research which have to be taken up by young researchers have also been identified and presented under the title, 'future research, besides identifying and recognizing the structure/process/stage elements in library and information studies[9].

Gaps in Literature

Keeping in view of the past scenario and current environmental location & situation, so far, a bibliometric study relating to Indian Geographical journals have not been attempted by any Indian Geographer on Indian landscape and it is considered to be a big gap. Hence, the present study has been taken up and it deals with the analysis of Indian geographical journals the author's pattern, subject content and change of geographic research literature from traditional to modern approach adoption, and application of Remote Sensing and GIS technologies additionally.

3. Objectives

The following are the objectives of this study:

- To identify and recognize the authorship patterns;
- To examine the number of papers that were contributed in the sub-fields of the science of geography;
- To adjudicate and adjudge the subject area of publication of papers:
- To prepare the quantitative assessment of geographic literature; and also
- To prepare the qualitative assessment of geographical research papers.

4. Material and Methods

The following are the sources of material:

1. National Geographic Journal of India, Varanasi.
2. National Seminar on sustainability of cities, environmental challenges,

   Department of Geography, Andhra University, Visakhapatnarn, 26-27. September, 2011.
3. Geographical Review of India, Department of Geography, Kolkata University. Kolkata, 20 I2.
4. Indian Journal of Regional Science, Kolkata, Vol. 44, No.1. 2012, pp.II-15.
5. Annals, National Association of Geographers India, Vol. 33, No. 2, December, 2013, p. 140.
6. Indian Journal of Information, Library and Society, Cuttack, Vol. 28, No. 3-4, July-December, 2015 and.
7. Indian Geographical Research: A Survey, R.K. Books, New Delhi, 2017, p. 205.

Method: To identify Current Trends (CT) in geographical research, Bibliometric Analysis has been applied. For details, the readers are requested to refer No.7 of the above list for cross verification.

5. Analysis

Different type of authored articles in six major branches of geography are presented in table-1 . It is seen from table-I, that there are four types of authored articles. They are (1) Single authored articles, (2) Two authored articles, (3) Three authored articles and (4) Multiple authored articles. The major branches surveyed are: (A) Physical Geography, (B) Economic Geography, (C) Population Geography, (D) Urban Geography, (E) Social Geography and F) Regional Geography.

Table 9.1: Different type authored Articles in Six branches of Geography

Sl. No.	*Type*	*Physical Geography*	*Economic Geography*	*Population Geography*	*Urban Geography*	*Social Geography*	*Regional Geography*	*Total No.or published papers*
1	Single Authored Articles	11 (44%)	7 (29.16%)	11 (44%)	10 (35.7 1%)	17 (53. 14%)	5 (17.85%)	71 · 43.82
2	Two Authored Articles	6 (24)	13 (54 .16)	12 (48 .00)	16 (57. 14)	13 (40 .62)	12 (42 .87)	72 44.44
3	Three Authored Articles	2 (8.00)	4 (16 .16)	2 (8 .00)	2 (8.00)	1 (3. 12)	1 (39.35)	12 (7.40)
4	Multiple Authored Articles	6 (24%)	-	-	-	I (3.12)	-	7 4.34
	Total No of Articles	25	24	25	28	32	28	162

It is further seen that (Vertical Analysis) in physical geography, single authored articles occupy the top ranking with 44 percent, two authored articles with 24 per cent, three authored articles occupy with 8 per cent and multiple authored articles contributes to a tune of 24 per cent. There are 25 research papers belong to this branch of geography.

(1) Gross Observations

1. Physical geography took the top list, and the single authored articles play an important and predominant role in current geographic research.
2. Three authored articles took the last position with 8 percent.

It is attributed that GIS and RS methods have been playing an important role in the present geographic research. Hence three authored papers are considered for publication.

B. In economic geography, two authored articles are at a high pedestal i.e with 54.16 percent, later single authored papers with 29.16 percent where as three authored articles with 16.16 percent and there are no multiple authors in the sub-field of geography.

(2)

1. It is attributed that two authored articles are enriched with GIS and RS Methodes"
2. Multiple authored articles are not at all published, if at all published, they are thematic with no GIS and RS methods The central theme with cartography enriched with aerial photo interpretation with critical map analysis is linked with topo sheet analysis.

In population geography sub-field, a total number of 25 papers are reviewed and 11 papers (44.00 percent) and 12 papers (48.00 percent) followed by two authored and single authored articles and three authored articles are to a tune of 2 with (8.00 percent). There are no multiple authored articles.

(3)

1. It is seen from the review that population geography sub-field is enriched with GIS and RS methodology, Previously this branch has enriched with mapping of variables and settlement pattern, with geometry followed by physical principles and with variations in social environment.

In urban geography, a total no. of 28 papers are reviewed. Among them, two authored articles have good place with 57.14 percent, where as single authored pattern followed by three authored articles with 36.71 percent and 8.00 percent respectively. Multiple authored articles are not at all published.

(4)

1. The technological applications are often find with two authored articles only.
2. Single authored and three authored articles are with traditional methods of data analysis and with the absence of GIS and RS methods. These papers are purely descriptive and no statistical methodology but these are good which are enriched with cartographic map analysis.

In social geography sub field 32 articles were published. Single authored articles were followed by two authored articles with 53.14 percent and 40.62 percent respectively. It is further seen that three authored and multiple authored articles are with 3.12 and 3.12 percent respectively.

(5)

1. It is observed that emigration, migration, travel analysis of the consumer and social mobility were linked up with statistical analysis, besides traditional map analysis. Social physics or social inertia is the major theme in social geography and social area analysis has been playing an important role in geographic research.
2. It is further observed that this sub-field is completely taking a new turn towards innovative social research methodology with scientific reasoning. GIS and RS methods are adopted and completely find new place in the place of qualitative analysis.

In the sub-field of Regional Geography, Regional Planning and Development Planning, a total of 28 papers were reviewed. Among them, two authored articles are high with 12 articles (42.85 per cent) whereas single authored papers were to a tune of 5 (17.85 per cent) followed by three authored articles to a tune of 3.57 percent. There are no multiple authored papers in this branch of geography.

In table-l, it is seen that a total no.of papers for each subfield has been indicated. In all subfields put together and two authored articles are 72 (44.44 percent) and single authored articles are to a tune of 71 (43.82 per cent. They find first and second rank and later followed by three authored and multiple authored articles were to a tune of 7.40 per cent and 4.34 per cent respectively.

(6)

1. It is observed that two authored papers tops the list with 44.44 per cent it and followed by single authored papers to a tune of 43.82 per cent first and second rank gained by two authored and single authored articles. It is attributed that single authored papers are considered as quite independent and two authored papers are considered as partly dependent for GIS and RS methodology and its application. In fact, it is a fact in deed.

Table 9.2: Typology of Articles (Loss/gain Analysis)

Sl. No.	*Type*	*No.of papers*	*Total No. of Papers*	*To be gained*
1.	Single Author Articles	71	71	0
2.	Two Authored Articles	72	72	72
3.	Three Authored Articles .	12 (3×12)	12	24
4.	Multiple Authored Articles	7 (7×4)	7	2
	Total	162	162	117

6. Typology of Articles (loss and gain analysis is presented in table-2

It is seen from table-2 that single authored papers, two authored papers, three authored papers and multiple authored papers are typical standing examples in the current geographic research. Two authored papers tops the list. Out of 162 papers, 72 papers belongs to this list. Three authored papers and multiple authored papers take the next rank respectively. They are 12 papers and 7 papers respectively.

Gross Observations

1. Single authored papers should have to be encouraged because 71 potential papers were published. There is no loss or no gain in this category.
2. Two authored papers category tops the list with 72 papers. For example 72 papers are distributed for 72 people. Another 72 people could not have presented 72 papers (72×2= 144-72=72) 72 papers were lost. If the two authored papers are not encouraged, these would have been published as single authored papers. It is attributed that original research capability should have to be encouraged in India on Indian landscape.
3. It is observed that three authored papers were only 12, 36 persons have contributed 12 papers. If three authored system papers are discouraged, then 24 people would be able to publish 24 more papers. The quantum of loss would be to a tune of 24. GIS and RS methodology could be developed independently than co-authorship.
4. It is further observed that four authored or multiple authored papers were only 7 (4×7=28). If the multiple authored papers are discouraged/abandoned. They would be able to produce another 21 papers.
5. It is observed that two authored, three authored and multiple authored papers would have been discouraged there is every possibility of getting potential papers to a tune of 117 instead of 162 papers during the reviewed period. The total papers would be (72+24+21=117) papers during 2011-2016.

7. Ranking of different authored papers

In table-3, ranking of different authored papers are presented single authored papers were assigned a weightage as 100 points, double authored papers were assigned weightage as 50 points. Three authored papers were assigned weightage as 25 points and multiple authored papers were assigned

12.5 points. Accordingly single authored papers were multiplied by its weight, then we get the net score. Physical geography gained. 1100 point score (11×100); Economic geography has gained 700 point score (7×100); population geography has gained 1100 point score (11×100); urban geography has gained 1000 points score (10×100); social geography (17×100) 1700 point score and finally regional geography has gained (5×100) 500 point score.

Table 9.3: Analysis Ranking of Different Authored papers

Sl. No.	*Name of the branch*	*A Single Author*	*B Double Author*	*C Three Author*	*D Multiple Author*	*Total (ABCD)*
1.	Physical Geography	1100 (18.03)	300 (8.33)	50 (16.66)	75 (86 .20)	1525 (15.11)
2.	Economic Geography	700 (11.47)	650 (18.05)	100 (33.38)	-	1450 (14.37)
3.	Population Geography	1100 (18.03)	600 (16.66)	50 (16.66)	-	1750 (17.34)
4.	Urban Geography	1000 (16.39)	800 (22 .22)	50 (16.66)	-	1850 (18.34)
5.	Social Geography	1700 (27.89)	650 (18.05)	25 (8.33)	12 (13.80)	23.82 (23.66)
6.	Regional Geography	500 (8.10)	600 (16.66	25 (8.3 3)	-	11 24 (11.18)
	Total	6100	3600	300	87	10087

At the second stage, two authored papers were multiplied by 50 point score, we have arrived at 300, 650, 600, 800, 650 and 600 point scores.

At the third stage, three authored total papers were multiplied by 25 point-score, we secured 50, 100, 50, 50, 25 and 25 total scores respectively.

At the final stage, multiple authored papers were multiplied by 12. 5 point scores we have arrived at 75 and 12 point score.

Findings

1. Single authored papers have secured 6, I00 point score. It is the top ranking authored papers with'A' .
2. Double authored papers have secured 3,600 point score, which got the second rank with 'B'.
3. Three authored papers have scored 300 point score and has gained third rank with'C' .

4. Finally, multiple authored papers got only 87 total score points and got 41 pt rank with 'D' in the total ranking system and all the scores were converted into percentages and they were indicated in brackets and all the four categories of papers were 162 and their score points were to a tune of 10,087.
5. Various sub-topics relating to geographical research were covered in all 162 papers.
6. Majority of papers with particular reference to geographical science relating to GIS and RS were published, traditional literature such as physical geography, economic geography, population geography, urban geography, social geography and regional geography including regional planning etc were finding place as independent thematic research.

10. Conclusion

This effort of the author may affect many students and new researchers to go deep into their subject and actively interact and interrelate the topic with interdisciplinary approach to strengthen their thesis to win their academic strength and make them ready for further reading of other papers of many authors. This study may indeed help thoroughly through the search in identifying the authorship pattern, subject branch that is more focused and sources of citation significantly preferred. Agenda for future researchers are preferably multi-disciplinary and inter-disciplinary in nature may find good placement. For ex. (1) land-form and land use studies (geomorphology and agriculture), (2) Eco-climatology, (geomorphology, ecology and climatology), (3) landforms and Oceanography (Geomorphology and Ocean based Land-from).

1. Resource management in a recently formed state (ie) Telangana state, India.
2. Exponential growth patterns of population and inter settlement patterns in India with special reference to Indo-gangetic plain.
3. Urban infrastructure and development policy studies such as (Urban economics, urban geography and urban sociology) with an interface of public administration finding a place to the local problems for achieving all round development.
4. Social dynamics and economic development in metropolitan cities like Mumbai, Hyderabad, Kolkata, Delhi and Chennai should have to be taken-up for structured research.

5. Development planning for an under-developed region such as a state/district/mandal or group of villages etc. and the theme should be coupled with demography, economics, sociology, political science, public administration, rural development and urban development in particular the policy science/policy studies and its impact on society which is to be deciphered.

Finally, it is concluded that interdisciplinary research and team spirit may help to occupy academic heights besides the application of technology with more quantitative methodology followed by core, such as Central traditional description in the thesis/monograph/project report is quite necessary. Additional technology may help additionally, without leaving traditional designing of the problem identification, recognition and finding solution to the selected problem both for intensive and extensive research stage by stage scientifically and efficiently which helps the society at large over a period of time, space, time and people are important elements in the regional system to improve on the line of development planning and finally be able to suggest a "model' for development.

Notes

* Consultant (Former). Regional Centre for Urban and Environmental Studies (RC UES), Osmania University. Hyderabad, Telangana State India.

1. Aadil Manzor Nadda and et.al., "Evaluation of Land uses land cover Dynamics in Vishav Watershed of Kashmir Valley (J&K)", National Geographical Journal of India, Vol. 60(4). December, 2014. pp. 381-390.
2. Ncch Singh. "Quality of Ground Water for Irrigation : A case study of a part of Ganga Basin". Geographical Review of India. 74(4), December, 2012, pp. 359-370.
3. Jay Karnakar Sanjay Kumar Mishra and M.S. Sukla Basu, "Issues of sustainability of City Slums: A case study". National Seminar on Sustainability Cities, environmental challenges. Department of Geography, Andhra University, Vishakapatnam, 26-27, September, 2011. p. 148.
4. Sayanatari Sarkar and Sumana Bandyopadhyay, "Proliferation of education and services in Peri-Urban regions of Kolkata: Perceptions of Providers and Beneficiaries". Indian Journal of Regional Sciences, Yo. 44. No. I. 2012. pp. 11-15. Jawaharalal Jain, "Mapping in India: An Evolutionary Perspective", Annual, NAGI. Vol. XXXIII. No. 2. December, 2013, p. 140.
6. V. Rangan & G.R. Parthasarathy. "Identification of Service Facilities of Urban Centres in Thene District, Tamil Nadu: A Geospatial Analysis. National Seminar on Sustainability of Cities: Environmental Challenges". Department of Geography, Andhra University. Visakhapatnam, 26-27 September, 2011. p. 153.
7. L.S.R.C.V. Ramesh and A.V.S.S. Naga Raju, "Publication Pattern in Indian Journal of Information. Library and Society 2007-2014: A Bibliometric Study", Vol. 18, No. 3-4, July - December, 2015. pp. 268-280.

8 Dr. M. Padmaja, "Bibliometric Analysis of the Journal of Rural Development: 2009-2013". Indian Journal of Information, Library and Society, Vol. 28. No. 3-4. July-December, 2015.

9 Simhachelam. J and Narasimha Murty, K.L. "Evolution and Evaluation of Research Methods in Library and Information Studies - Bibliometric Study: A Survey (2005-2009), published in Indian Journal of Information, Library and Society I.JILIS). Cuttack. Vol. 26. No. 3-4, July - December, 2013, pp. 174-192.

The author is indebited to Gram Panchayath Officer, Mandal Revenue Officer, Agricultural extension Officer and statistical assistant for providing necessary data for this research activity and also to focus group members in the field. Further, the author expresses his gratitude to various persons, namely Sri Akella Kama Sastry Garu (Head Master, Retd) Sri Kesiraju Satyanarayana, Sri Tallaprapragada Gopalakrishna, Homoeo Doctor, Sri Kaigala Venkatachalam, Sri Shaik Masthan, Sri Shaik Sayed Bagi, Sri Kanuri Satayanarayana, M.A., B.Ed., (Retd.) Sri N.V. Subrahmanayam M.E.O. (Retd.), Sri Madhavarapu Venkata Rao, HM (Retd.), Sri Konala Narayana Rao, Sri Tylem Atchuta Rama Rao and Sri N. Peri Sastry, HM (Retd.) have helped to the maximum possible extent in this endeavour. Sri Chamarti Lakshmipathi has helped and responded to the author for this research activity to a large extent.

Research Paper Two

Resource Base and Pattern of Development A Case Study of Vegeswarapuram Village of Tallapudi Mandal, West Godavari District, Andhra Pradesh, India

K. L. Narasimha Murthy

Abstract

Resources are sources of energy for development and helps for the alleviation of human - economic stabilization, sustainability and Susceptibility water helps to change the agricultural land use and crop potential patterns (causative and effective factor) besides social transformation from social stratification and cultural functions. Physical process includes location, physical setting, climate, temperature, rainfall and soils are analysed. Classification of economic resources - land, landuse and cropping patterns may change due to source of irrigation and its intensive activity over a period of time has also presented. Land use and population, migration of population for employment (outmigration), and employment scenario of incoming people (irrigation) process at the village level (Vegeswarapuram) and identification of infrastructural facilities (over heads - social) and cultural factors that helps to co-ordinate people (aggregation of people, tradition and so on) are examined in detail at the lowest level of administration. Government of India and Government of Andhra Pradesh's policies and programmes that have been operating/operated in this village for rural development are particularly examined, adjudged and adjudicated. Micro level unit of administration, formation of mandal and its head quarters, whether they are properly located or not is an important factor and suggested new location as the mandal headquarters is truly Signifies the justice or not, keeping in view of the geographic viabilities and vulnerability. The aforesaid factors have been examined critically. Finally; at the end, the author has made an attempt by identifying the four processes and its operation in a deltaic village and the settlement system.

1. Introduction

Resources are sources of energy and helps for the alleviation of human culture and rural development. The author presents the resource base and pattern of development over a period of time (1975 and 2017) both years inclusive time and space study of Vegeswarapuram village of Tallapudi Mandal, West Godavary District, Andhra Pradesh, India (Mandals of West Godavari District, location map of Vegeswarapuram village, list of villages of Tallapudi MandaI and MandaI performance have been presented - reference - Annexures) broadly, this paper deals with the following aspects and they are:

1. Physical process (pp)
2. Human economic stabilization (H.E.S.)
3. Social transformation and Social stratification (S.T. and S.S)
4. Cultural formation (New Culture) (C.F.).

1. Physical process includes location, physical setting, climate - temperature, rainfall and soils.
2. Classification of economic resources - land, land use and cropping patterns (Irrigation and its activity).
3. Land use and population - Migration - outmigration and immigration process at the micro level unit - Vegeswarapuram.
4. Infrastructural facilities (Social overheads).
5. Cultural Aspects:

   "Increasing demand for human sustainability within the framework of decreasing supply of facilities has been assessed and also presented the impact assessment (I-A) upon land and people" keeping in view of the government programmes and its augmentation for rural development have a telling effect in Vegeswarapuram village and its assessment has been made besides mandaI formation, development and implementation of MandaI policy of Sri N.T. Rama Rao, the then Chief Minister of Andhra Pradesh and also Sri N. Chandra Babu Naidu, the present Chief Minister of Andhra Pradesh state and impact of a Janmabhoomi" programme at the village level.

2. Literature Survey

Rural development studies covers the themes such as local resources (indigeneous) helps for rural development besides grassroots perspective.

Non-farm sector also contributes to rural development. Decentralized governance, is the latest theme involved in rural development studies and 21st centuary has made a new way for the furtherance of development. Sustainable development and MGNREGS are the two important programmes and policies of Government of India, helps to alleviate rural poor and stepup people to bring them above the povertyline. Krishna Mohans addressed the regional backwardness and presented the area development programmes in India (2005) and Nath 2, has identified that planning and development are the two sides of a coin, further stressed the need of planning for development (2010). Anita Kumari 3, has presented the issues and policies related to balanced regional development in India. Srivastav, *D.S.* 4, has high-lighted the necessity of regional planning for an economic region. 'Regional planning' helps for sustainable development as envisaged by Shukla (Amitabh) 5, (2012). Chauhan 6, has identified two themes in his study i.e, Economic Agriculture and Population (P.R) Growth (2014). Ruth S.K. 7, in his macro study has further identified that landuse change and agricultural growth in India are two impartant causative factors for rural development. Ferroni, Marco 8, has presented the natural scenario that is transforming Indian agriculture, India, 2040 (2013) and Singh, It has presented the integrated Area Development and planning (1998). Murthy, K.L.N. 10, has presented the Integrated Area Development Plan for Tanuku Taluk, West Godavary, Andhra Pradesh, India (1985).

2.1 Gaps in Literature

It has been identified and recognized that no micro region has attracted geographers to present the problems and challenges relating to a village and find solution to it till 1970.

2.2. Problem

Resource Base and pattern of Development in a village has not been covered and it is still an uncovered problem. The author opined that integrated resource survey and rural development - A case study of Vegeswarapuram of Tallapudi MandaI, West Godavari District, Andhra Pradesh, India has been surveyed (First Survey during 1975) and the second survey has been undertaken (2016 and 2017) in order to present the process of programmes of Government of India and Panchayat Raj Department of Andhra Pradesh and its impact on Vegeswarapuram village and further present the results in this paper.

2.3 Objectives

1. To identify the resources for rural development.
2. To classify the quality land (yield per acre).
3. To classify economic factors - Well irrigation and canal irrigation and also present Hydraulic particulars of A.P. irrigation Development Corporation Limited, Vegeswarapuram pumping scheme, Kovvur Taluk, West Godavari District, Andhra Pradesh.
4. To identify the general land use pattern and irrigation in Rabi Season, 1975-76 and land utilization - 2016 and cropping pattern 2016.
5. To recognize land use and population - size of fields and farmers.
6. To identify and recognize infrastructural facilities (Social over heads) and also examine the central functions, (i.e.) education, banking, health, veterinary, educational facility, industry, agro-based commercial centre, communication, land registration and cultural centers, temples, churches and mosques etc. for the overall development process.
7. To identify the Development pattern as on 31-12-2017 (A comparative Analysis).
8. To identify the environmental impact assessment (EIA) in Vegeswarapuram village and cultural events.
9. To identify and recognize the migration processes.
10. To measure the social stratification patterns, the social interaction process includes changing ownership of houses, house-sites from one community to the other and also decipher the vanishing community at the gross-roots level.
11. Cultural centres - Rural Development and also
12. To list gross observations

3. Physical Factors

3.1 Location:

'VEGESWARAPURAM' is a village in Thallapudi Mandal in West Godavari District of Andhra Pradesh State, India. It belongs to Andhra region. It is located 87 k.m. towards East from District head quarters Eluru, 3 km from Tallapudi - 397 km from State capital Hyderabad. Vegeswarapuram pin code is 534 340 and postal head office is Vegeswarapurm.

Tallapudi (2 km), Chidipi (2 km), Prakkilanka (2 km) Peddavam (4 km), Gajjaram (4 km) are the nearby villages to Vegeswarapuram. Vegeswarapuram is surrounded by Seetanagaram MandaI towards North, Kovvur MandaI towards South, Gopalapuram MandaI towards West, Chagallu MandaI towards South. Kovvur, Rajahmundry, Nidavavole & Tadepalligudem are the nearby Urban centres to Vegeswarapuram. This place is in the boarder of the West Godavari District and East Godavari District. In East Godavari District, Seethanagaram is North towards this place. The latitude 17.4 and longitude 78.48 are the geocordinate of the Vegeswarapuram and this village is a Panchayat. The native language of Vegeswarapuram is Telugu, Urdu, English and most of the village people speak Telugu, Urdu, English.Vegeswarapurm people use Telugu, Urdu and English languages for communication. The following Vegeswarapuram map is from Google, Vegeswarapuram map consist of nearest villages and core information is based on zooming. Total persons are 52,785 total population male 26,295) and total population female are 26,490 and in Tallapudi no. of households are 15,109 and there is no urban population. The area of the village Vegeswarapuram is to a tune of 788 hectares. List of villages in Thallapudi MandaI are as follows. They are (1) Thadipudi (2) Ragolapalle (3) Thupakulagudem (4) Pochavaram (5) Paidimetta (6) Prakkilanka (7) Gajjaram (8) Annadevarapeta (9) Kukunuru (10) Thallapudi (11) Vegeswarapuram (12) Ballipadu (13) Peddevam (14) Tirugudumetta (15) Malakapalli and (16) Ravulapadu (Reference: West Godavari - Mandal Tahsil Map, Google Maps and Vegeswarapuram village Maps are presented. Annexure - 1 shows the 16 villages census structure and composition of Population with segmented pattern have also been presented.

The historical name of the village was "Yogeswarapuram' which denotes notably well noted centre for Vedic culture. Among the vedic pandits) Sri Late Kalluri Veerappa Somayajulu, Krishna Yajurveda Ghanapati, Urlam Samsthan and Kasi (Varanasi) Pandit has performed 'Yajnam' in the year 1900 A.D. and Sri late Kalluri Ramabhadra Chaynulu, Dharmasastra Chakravarthi and Apara Apastambulu were the natives of this village.

Demo-graphics of Vegeswarapuram.

Telugu is the local language here. The population of Vegeswarapuram is 5,552. Males are 2,845 and females are 2,707 living in 1392 houses. Total area of Vegeswarapuram is 788 hectares. Kovvur is the nearest town to Vegeswarapuram. Kovvuru is 14 km from Vegeswarapuram. Road connectivity is there from Kovvur to Vegeswarapuram.

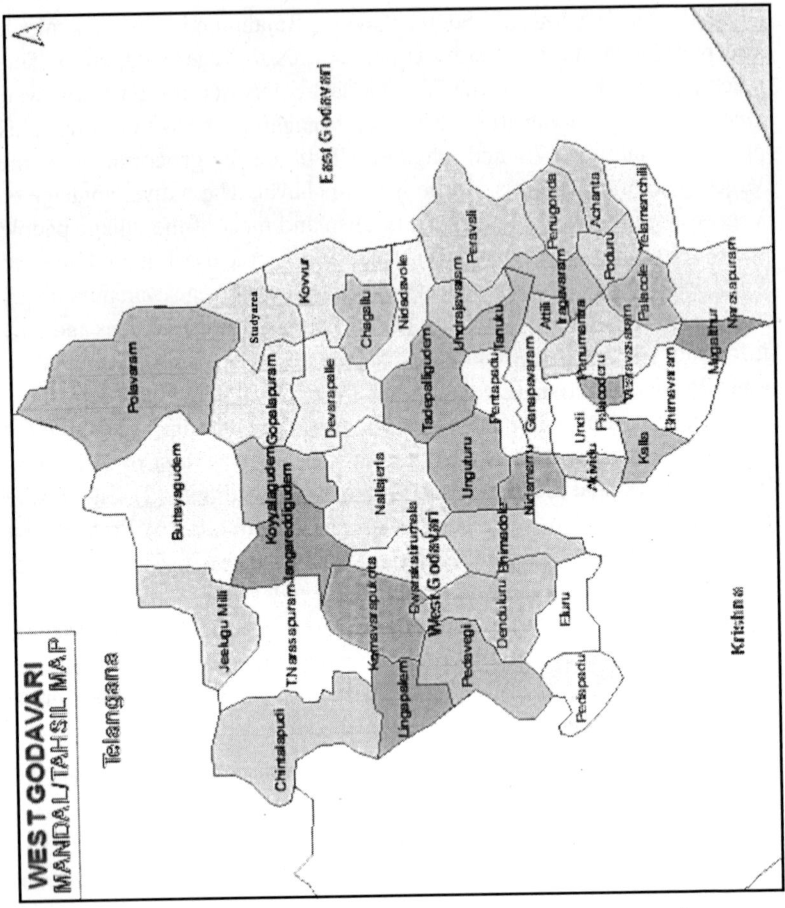
WEST GODAVARI
MANDAL/TAHSIL MAP
Telangana
East Godavari
Krishna
West Godavari
Polavaram
Buttayagudem
Jeelugu Milli
Chintalapudi
T.Narasapuram
Koyyalagudem
Gopalapuram
Devarapalle
Kovvur
Chagallu
Nidadavole
Tadepalligudem
Nallajerla
Kamavarapukota
Lingapalem
Pedavegi
Denduluru
Eluru
Pedapadu
Bhimadole
Unguturu
Nidamarru
Ganapavaram
Pentapadu
Tanuku
Undrajavaram
Peravali
Penugonda
Achanta
Attili
Iragavaram
Penumantra
Poduru
Palacole
Yelamanchili
Narasapuram
Mogalthur
Bhimavaram
Kalla
Akividu
Undi
Palakoderu
Veeravasaram

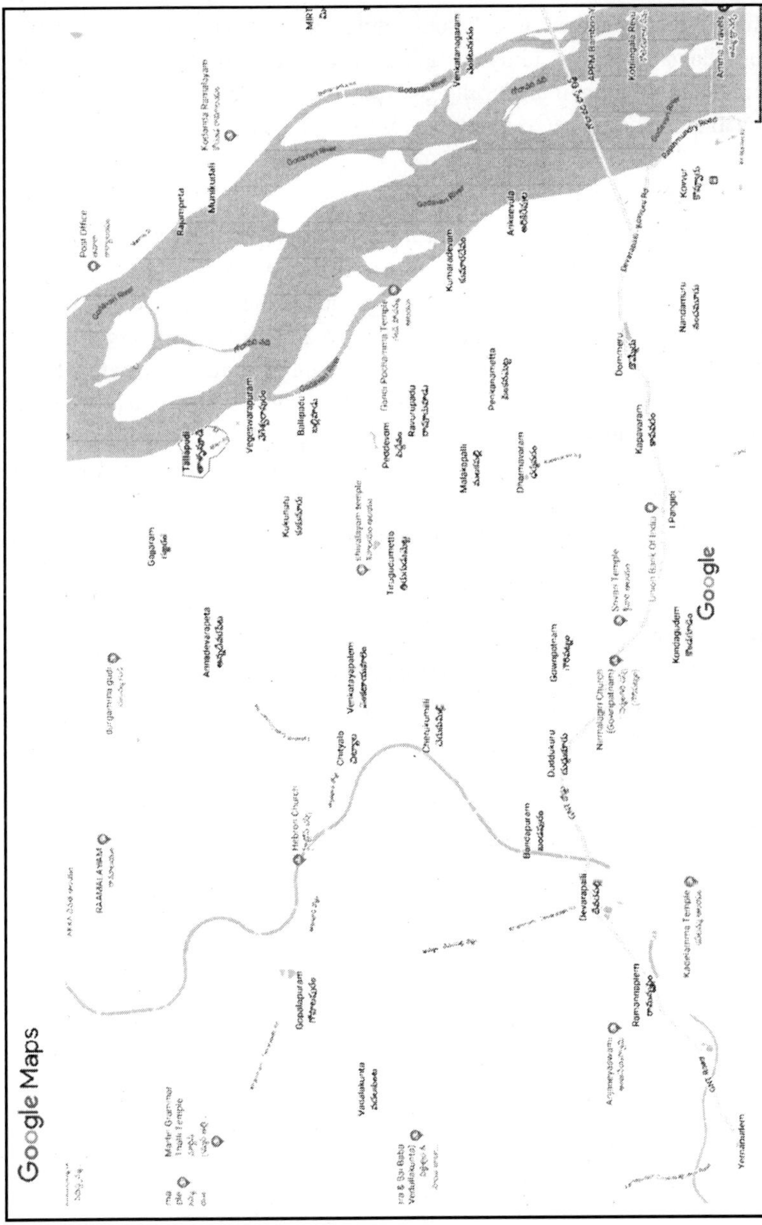
Google Maps
Google
Godavari River

The area under study is a part of daltaic plain, the land was cultivated under rainfed conditions (1975) and the depth of the water is almost 20 meters, at the most under which there is a hard layer of granite. The general slope of the village is towards south and south west and the water table in the village during the rainy seesaon is at the depth of 2 meters, while in winter and summer, it drastically goes down to 20 to 25 meters respectively. The average bed level of the village is 19.60 meters.

3.2 Climate

Climate is regarded as a controlling factor of land use, as it influences soil, vegetation, agricultural land use, agricultural crop patterns and other humn activities. No climate data are recorded in the village, so Kovvuru is the nearest meteorological station, where weather information is collected and the same has been taken as close approximation of data of the Vegeswarapuram village. The highest rainfall was 1122.80 *mm* (1975) and the lowest being 802.80 mm. It has been observed that the maximum and the minimum difference is to a tune of 200 mm the recorded highest temperature is 45°C the lowest with 40°C during the summer season. Summer month is highly difficult for humans even though it is located in a river prone environment.

3.3 Floods

Due to the presence of high rainfall, floods are occured and this village has been experiencing the floods with a good frequency. Public works Department (PWD) officials have been taking certain measures as precaution for relief from floods. The Godavary flood waters recorded to an extent of 17.50 feet, which was the mean flood level (M.F.L.) during 1953 and 1975. The water level has touched the danger mark. This is the maximum height, naturally has been touched and recorded during the flood season.

Table - 1 Flood level

ISI.No.	Date	Flood Level (F.L) in meters
1.	1-7-1953	22.00
2.	2-7-1975	20.95
3.	16-8-1975	21.29
4.	10-9-1975	21.75
5.	11-9-1975	21.77
6.	9-9-1999	21.80
7.	8-9-2015	21.00

Source: P.W.D., Vegeswarapuram (Field work)

It is seen from Table 1, the flood level was highest on I" July, 1953 to a tune of 22.00 m and the later two decades, the level has been decreasing from 21.80 m to 21.00 meters. The impact of floods during 1953 was more than the remaining years.

3.4 Soils

The area under study has no detailed soil survey from textural point of view and the department of agriculture has prepared the soil map according to its fertility. The soil of this village comes under black (deep) and the deltaic soil (clay, sand and silt) leads to alluvial soil. Hardness, less aeration and water on the surface which is bad for percolation of water and difficult to tillage are the typical characteristerics of the black and black regular soil. The addition of organic and chemical fertilizers in large quantites have been improving its texture. It is also observed that additional inputs leads to additional output to the average yield of the crop.

4. Economic Factors

Physical factors, among them, rainfall, temperature, precipitation and soil structures may decide the typical land use and cropping pattern. Besides physical factors, economic inputs may also to some extent, decide; the cropping pattern and structural components of technology. The general terms, as per the renowned economist Sir Alfred Marshall has identified that technology play an important role in agricultural crop productivity. Physical factors, economic factors and technological factors also play an important role in the agricultural crop productivity i.e. crop yield per acre in kilograms, the farmer receives during that crop period is a deterministic factor. The soils may decide the cropping pattern especially in rainfed conditions.

4.1 Classification of land according to fertility

The strength and stability or soil potential could be measured in terms of crop yield (crop potential), in other words, we may call on the basis of fertility, the land may be classified according to the following scheme.

Table -2

SI.No.	Classification of land	Paddy per acre Yield in bags
1.	Good quality land	45
2.	Medium quality land	35
3.	Low quality land	20
	Total	**100**

It is seen from Table 2, the quality of land has its impact on crop productivity. The good quality land gives 45 bags of paddy per acre has been presented and the quality of land determines the yield of crop per acre. Almost the quality of soil has its impact on the quantity of yield, which differs from quality and also from land to land and from area to area (area differentiation to a large extent).

4.2 Economic Factors

"Irrigation is an influencing factor for cultivating the land as well as crop culture to the optimum level and its potential will be stepped-up from one generation to another generation", helps for the social and economic transformation from social stratification" and from economic drought to agri-efficiency" (Quotation mine please). The village has ample irrigation facilities as compared to the other parts of the surrounding areas and regions. There are two principal means of irrigation sources. They are: (1) Well irrigation and (2) Canal irrigation.

(i) Well irrigation:

The village has the facility of these masonary wells for irrigation and they are distributed at a distance. They supply the water throughout the year to the fields, is an important factor for good agriculture. Before the construction of the Godavary river Pumping scheme, most of the area was totally dependent on rains and rainfed crop culture was much visible and the vissitudes are quite rampant, recurrent and repetitive".

(ii) Canal irrigation:

The Godavari river water has been diverted from the main source into canal which enters the village from East, North and West and further flows towards south and south-west direction. The canal is 5 km length and it supplies water during the 11 months for irrigation in a year, under which, both kharif and rabi crops are sown.

Table 3: Andhra Pradesh Irrigation Development Corporation Limited Vegeswarapuram Pumping Scheme, Kovvur Taluk, West Godavari District Main Canal Hydoraulic Particulars

1.	**Length**	**5.00 km**	
2.	Total Ayacut 6000	3,000 wet	3,000 dry
3.	Discharge	Head 2.44 cusecs	End 0.45 cusecs
4.	Duty	Wet 60	Dry 120
5.	Suction	Head 4.9 m × 1005 m	End 20×0.9 m

(Contd...)

6.	Bed level	Head + 19.60 m	End + 18.60 m
7.	Full supply level	Head + 20.65 m	End + 19.50 m
8.	Top Bank level	Head + 21.40 m	End + 20.25 m
9.	Surface fall	115,000	
10.	Top width of Bank	Head 2.0 m	End 1.5 m
11.	Slopes	Front 1:1	Rear 2:1

Data have been collected from Asst. Engineer, APS.IR.D.C., Vegeswarapuram Pumping Scheme (Author's field work).

4.3 The general land use pattern in the village in 1975 is presented in Table - 4.

Table 4: Land Utilization

Sl.No.	Land use	Area in acres	Per cent to the Total area
A.	Cultivable land	1722.96	88.67
1.	Cultivated land	1500.00	77.20
2.	Gold fallows	34.64	1.78
3.	Current fallows	40.00	2.05
B.	Non-Cultivated land	74.40	3.82
	Settlement		
	Village paten,		
	Tank		
	Public Wells		
	Public Offices		
	Edn. Institutions		
	Temples & Churches		
	Varginia Tobacco company		
	Waste-Land		
	Other cultivated land	73.96	3.83
	Other than fallows		
	Others	220.14	11.32
	Grand Total	1943.14	100.00

Data has been collected by the author (field work) from Gram Panchayat office, Vegeswarapuram, during (September - December, 2017).

It is seen from Table 4, land utilization statistics, that cultivable land is to a tune of 1722.96 acres, which accounts for 88.67 percent. The cultivated land has occupied by 1500 acres, which accounts for 77.20 per cent. The old fallows has been occupied by 34.64 acres which accounts for 1.78 per cent. The non-cultivable area has been occupied by

74.40 acres (3.82 percent) others occupy to a tune of 11.32 per cent other cultivated land and other than fallows is to a tune of 73.96 acres (3.83 per cent). The total area is to a tune of 1943.14 acres.

4.4 Land Utilization in kharif season, 1975

The land devoted to kharif crop in 1975 is only 88 acres (5.8 per cent) out of the total cultivated land and the whole of this land has been devoted exclusively to sugarcare production. The remaining cultivated land was left fallow for Rabi crops. The sugarcane acreage is also very small because of the insufficient water supply for the whole survey numbers. This has been improved by bore wells, which were located in the fields at 40 to 50 feet depth, where plenty of water is available. These fields are located near the river bund.

4.5 Land utilization in Rabi Season, 1975-76

The area occupied by each crop in this season is considerable and is shown clearly that out of gross-cultivated land of 1520 acres. Rabi crops were grown on 1432 acres (94.2%) of the land remaining 88 acres (5.8 per cent) of land was occupied by continual kharif crop, leaving the fallow land. It is clear by the evidence that the only important crop to this village is Rabi (Ref. Table 5).

It is quite clear that out of the total land of 1943.10 acres, 1720 acres of land (88.5 per cent) is cultivable land. Out of this category, cultivable land (18 per cent and 2.5 percent) land is old and current fallow respectively. 74.75 acreas (3.8 per cent) is at present placed under non-cultivable land and other uncultivated land other than fallows are only (3.8 per cent) and the remaining (77.20 per cent) land has been cultivated.

Table 5: Irrigation in Rabi Season 1975-76

Sl.No.	Name of the crop	Area in acres	%of gross cultivated land	Total % of gross cultivated land	Total % of Net cropped land
1.	Grain crops	60	3.9	11.2	94.21
	1. Redgram				
	2. Dhania	30	2.0	—	—
	3. Jowar	15	1.0	—	—
	4. Maize	10	0.7	—	—
	Total	**115**			

(Contd...)

2.	Cash crops	22	1.5	—	—
	5 Varginia				
	Tobacco				
3.	Commercial crops	1150	75.60	83.0	—
	6 Chillies				
	7 Others	145	9.50		
	8 Continual kharif Crop (sugarcane)	88	5.80	5.8	5.79
	Total	**1405**	**100.00**	**100.00**	**100.00**
	Total	**1520**			

Source: Data has been collected from field by the author during 1975-76.

Table 6: Land Utilization - 2016

Sl.No.	Land use	Area in acres	Percent to the Total area
A.	Cultivable land	1720.00	
	1. Cultivated land	1400.00	81.39
	2. Old fallow	34.00	1.97
	3. Current fallow	100.00	5.84
B.	Non-Cultivated land	120.00	6.97
	Settlement		
	Village path		
	Tank		
	Public Wells		
	Public Offices		
	Edn. Institutions		
	Temples, Churches & Mosques		
	Varginia Tobacco company		
	Waste Land		
	Other cultivated land	66.00	3.83
	Other than fallows		
	Others	1720.00	100.00
	Grand Total	1720.00	100.00

Source: Authors field work during September - December, 2017.

4.6 It is seen from Table 6, the land utilization statistics data that the cultivated land was to a tune of 1400 acres, accounts for 81.39 per cent, where as old fallows and current fallows land area to a tune of 1.97 and 5.84 per cent. The non-cultivated land is to a tune of 6.97 per cent, and other cultivated and other than fallows are to a tune of 3.83 per cent.

Table 7: Cropping pattern - 2017

Sl.No.	Name of the crop	Area in acres	Percent to total cropped area
1.	Paddy	750	53.57
2.	Sugarcane	500	35.71
3.	Plantain/Banana	150	10.72
	Total	**1400**	**100.00**

Source: Focus group discussion Data collected through interviews by the author during fieldwork September - December, 2017.

4.7 Cropping pattern

The cropping pattern relating to 2017 has been presented in Table 7. It is seen from the table 7, that the paddy, sugarcane and plantain / Banana crops are important crops. The paddy crop occupies nearly 53.57 per cent area to total cultivated land, sugarcane is to a tune of 35.71 per cent besides banana crop occupies nearly 10.72 per cent of land. These three crops are considered as important mono-cultural crops and rabi crops are not visible as plenty of water is made available for agricultural use / operations.

During the year 1975, many crops were sown and has been identified the following crops. The important crop combinations were as follows:

(i) Paddy + red gram (ii) paddy + til and Kharif crop (continual) sugarcane. The first two category crops are not at all visible and vulnerable and this wet type of farming will not sustain other dry crop patterns.

4.8 Land use and population

In ‘Vegeswarapuram’ village the land has been used optimally. The total land that is available for cultivation is 1400 acres. The total population of the village is 5601 (latest figure as on date). The total number of households are 1598. Main cultivator population is 121. The average size being to a tune of 11.57 (1400/121=11.57) acres. (Reference Annexure-I).

Gross observation

The available land for cultivation has been reduced by 200 acres, as this land is being used for settlement purpose by disturbing the mango garden, which is nearby the Kovvur - Tallapudi Main road, at the entrance of the Vegeswarapuram village used for settlement over 1975 (it is a major change in land use (Refer: Table 7 also).

4.9 Size of fields and the farmers

It is observed that the fields sizes are varied and vivid. The largest size of field is to a tune of 20 acres and the smallest size of field is 0.75 acres number of non-working population is 3269. Persons, non-working male population occupies to a tune of 1067 and non working female population is to a tune of 2202 persons. Females outweigh male population (See Annexure-I).

It is further observed that 58.36 per cent of population belongs to non-working category and 41.64 per cent are working population. It is assessed that 41 per cent population are working population and agricultural operation category occupies to a tune of 30 per cent. Hence 30 per cent people are engaged in agriculture.

5. Infra Factors

The quality of life of people depends upon the quantity of inputs that have put in for operation. The infrastructural facilities are classified into the following and they help for 'integrated' area development. The facilities help for the multi-dimensional and multi sectoral development at the village level. The facilities are as follows, which helps for rural development process for the total development at the village level. They are:

1. Education 2. Banking 3. Health 4. Veterinary 5. Maternity 6. Transport and 7. Cultural.

5.1 Education (Reference: Google Map-3).

1. K.S.R.Z.P. High School, previously it was known as K.S.R.B.H. School and it was established during the year 1947 and had celebrated the 70th Anniversary on 8th and 9th July, 2017. Now, the name of the school has been changed as K.S.R. Government High School, Vegeswarapuram.

2. Christian Missionary School offers education upto 8th class since, 1948.

3. Government Junior College, Vegeswarapuram was started in the year 1979.

4. P.S. Jr. Basic School, Vegeswarapuram.

5.2 Banking

Primary Co-operative Society Bank (PCSB) Vegeswarapuram.

5.3. Health

Maternity and Child Welfare centre.

5.4 Veterinary

Animal Husbandry Centre (A.H.C.)

5.5 Continuing Educational facility

P.S. Library, Vegeswarapuram (formerly in Gopalapuram Block)

5.6 Industry - Agro based

1. The Upper Godavary Khandaseri Sugar Mills Ltd. Kukunoor.
2. Two Rice Mills (Agro.based)
3. Three Saw Mills (Forest based)

5.7 The Agro-based Commercial Centre

The Indian Leaf Tobacco (I.L.T.D) - Purchasing centre

5.8 Transport

1. APSRTC - Identified Bus stop
2. APSRTC - 2 Request stops

6. Cultural Centres

1. Six temples
2. One Church
3. The Godavary Mary Mata Temple
4. One Mosque

(Reference Google Map - Geographic location of resources).

7. Communication

1. Telephone Exchange (BSNL).
2. Sub-Post Office
3. BSNL Office

8. Land Registration

Sub..Registrars Office, Vegeswarapuram

Google
Google Maps

9. Artisians

1. Idol making - from the caste of Golla
2. Honey collection - S.T. Yanadis
3. Wood work based on machinery (wood work) specialist in wood work.
4. Potters - Wheel
5. Medari - Special craft. basket makers)

Multi-dimentional and Multi-sectoral Development (Changes over 1975 and as on date 31-12-2017)

1. It has been observed that 'Vegeswarapuram' village has been developing in multiple demensions he integrated rural development. Different sectors are as follows:

1. Education 2. Banking 3. Health 4. Veterinary 5. Library 6. Industry 7. Agrobased commercial centre 8. Cultural centres.

1. Government Junior College has been established in the same educational complex of KSRZPH School and a separate building has been constructed as Government Junior College, Vegeswarapuram, to cater to the needs of the people in an area of 10 km radius.
2. KSRZPH School has been taken over by Andhra Pradesh Government and now they have changed the name as KSR Government High School, Vegeswarapuram.
3. MandaI Parishad Adarsha Pathasala (Main), Vegeswarapuram, Tallapudi Mandal, Sarva Sikshabivan was established since 8 years, offers Upper Primary Classes and Primary classes from (2007) with RC No.193/48, dt. 7-5-1948.
4. Sri Sarswathi Vidyaniketan, Vegeswarapuram has been offering 1st and 2nd classes, which was started in the year 1988, by an women entrepreneur of the same village and this institution needs recognition by A.P. Government.

2. Continuing Educational facility - P.S. Library

The Panchayat Samithi Library building has been collapsed and now, this facility has been lost for ever, as there is no person to try and regain this facility.

3. Co-operative Society

The Primary Agricultural Co-operative Society, Vegeswarapuram, has its own building, and it was declared opened by Sri Yerra Narayana Swamy, M.A., L.L.B.,

Chairman, Zilla Parishad, West Godavari District, Andhra Pradesh on 30th December, 1987. Sri Kurasala Narasanna (Pedda kapu) was Chairman of the Bank. Later, this facility has been shifted to Annadevrapeta (another settlement) due to changed political scenario.

4. Veterenary Hospital (V.H.)

Under Janma Bhoomi programme, Valagapudi Ramakrishna Veterinary

Hospital, Vegeswarapuram has been constructed during the year 1999 and this was declared opened by Smt. Rajeswari Rama Krishanan, Managing Director, the Jayposre Sugar, SChagallu, West Godavary District, Andhra Pradesh.)

5. Revenue Office

Now, Vegeswarapuram, Grama Revenue Office has also been (taking) functioning in the same building premises since a longtime.

6. Health Sub-Centre, Vegeswarapuram is being run by Smt A. Vijaya Kumari and Padmavati, A.N.M., Pradhanamantri Surekishat Matritva Abhiyan (PMSMA) is functioning on good-lines which caters to the needs of the people.

7. General Health

1. Homoeo Doctors- 2 (DHMS)
2. Indian Medicine - Registered Medical Practioner (RMP)-1.

6. Social Stratification and Social Interaction Process At Micro Level Settlement, Vegeswarapuram.

The social area analysis leads to an important aspect, which leads to a ray forward, keeping in view of the past, present and future scenerios in any state or nation or at the international level. Generally and specially also a notable note is to be examined in this village settlement system is as follows.

Geographically from the North-East of Vegeswarapuram village, the mighty river Godavari water is entering through APSIRDC,

Vegeswarapuram pumping scheme has started in the year 1974. The perennial Godavari Waters have changed the resource base totally and a new pattern of development has been developed since 1974. The pumped water has been entering from the East-North East and further from east to west the canal has been dug, constructed and taken a 'U' turn after passing a length of 2 km and further taken a turn towards, West-South West direction to cater to the needs of the people. It has been observed that Godavari water totally changed the irrigation system from dry to wet pattern. The physical processes leads to the economic stabilization of the farmer to a large extent, lateron, the economy has also been stabilized inclusive of the farmer to strengthen financially and also empower to create landbase, property acquirement of lands, housesites, development of sites and also new house construction activity of r.c c buildings to a tune of 500 houses.

It has been observed that on the southern and south East fringe of this village (keeping in view of the past vedic culture) the S.E. population has settled and constructed the thatched houses and also with Bangalore tails of settlement system. In the South-west direction the mighty church is located. On the eastern fringe among the cultural centres, 3 temples are located and further west, the Yanadi colony / maternity centre is located. Because of the dearth of space, at present, on the south and southwestern side, the Yanadi (Scheduled tribe population - Reference Annexure-I) has been established as an extension, nearby, the upper caste people, such as naidu's houses are also located along the lines and rows. It shows that the society has been changing and it is a sign of social acceptability and social interaction with the upper category of the social lader.

In the social ladder, step by step interaction has been taking place mainly in the main street of Vegeswarapuram, (previously the Brahmins who have resided in the main street) High School street because, the traditional old 'Vedic culture' is visible in this street. It has been observed that over a period of time, all these Brahmin families have been migrated for employment, as education enabled them to secure some employment in those days from this village to the national capital region (NCR - New Delhi and State capital twin cities of Hyderabad and Secunderabad. As a result, the Brahmin community sold the agricultural lands and also house property to the Naidu families. Here, in migration particularly out migration is visible to a large extent.

It is further observed that the post pumping scheme of Vegeswarapuram Godavari Water pumping scheme changed the irrigation system and thus, resource base has left a new pattern of development. It is proved that irrigation system has changed not only the crop and crop patterns but also land use and land utilization includes gaining of economic strength and attained political maturity has changed the traditional old president to a new person to assume to power. As a result, Sri Kommireddi Parasuram has becomes sarpanh of this village during 1990-2000. These two powers enable them to purchase lands, houses and house sites by Naidu families from Brahmin families and left a new settlement system into a town cultural relation (Reference: Changing ownership of houses, house sites and agricultural lands from the Brahmins by Naidu families).

Table 8: Changing Ownership of houses and house sites from Brahmins to Naidu families

Sl.No.	Name of the Owner	Year
1.	Sri K.R.B. Chaynulu (H) Sri Ch. Seetiah	1982
2.	Sri K.R.B. Chaynulu (H.S.) Sri Chodisetti Perraju	1977
3.	Sri Y.V.S. Murthy (S) Sri V. Satyanarayana	1982
4.	Sri Akella Venkataratnam & Sri Akella Sesha Rao (H)	No change has been registered
5.	Smt. Yeluripati Chinna Venkamma (H) Sri Matireddi Appa Rao	1978
6.	Sri Nori Narasimha Sastry (H) Sri Petta Krishna Murthy	1985
7.	Sri Ch. Rama Bhadra Sastry (H) Sri R. Pitchaiah	1978
8.	Sri Kallakuri Umamaheswara Rao (H) Occupation Unauthorised	1990
9.	Sri Malladi Srirama Sarma Sri M. Ramakrishna Rao. (H) Sri M. Suryanarayana	1991
10.	Sri Ch. G.S. Murthy (H) Indugula Basayya	2016

Source: Authors field work (October - December, 2017) Sub Registrar' s Office, Vegeswarapuram.

Population Migration (outmigration and immigration) has taken place in this village due to employment other than the native place. Migration for employment is a causative factor of sign of temporary migration later on, permanent migration has taken place where they are employed.

Table 9: Outmigration and immigration process (From Vegeswarapuram to another important places) (See Annexure - 1)

Sl.No.	Number of people Immigrated (Caste-wise)	No.	Year	Place
1.	Brahmins	50	1950-2017	Hyderabad & Vizag
2.	Naidus	8		Hyderabad
3.	Padmasali	8		Hyderabad
4.	Rajaka	2		Polavaram and Jangareddigudem
5.	Christians	4		Visakhapatnam
6.	Muslims	1		Rajahmundry
7.	Others	27		Visakhapatnam & Rajahmundry
	Grand Total	100		

Source: Author's field work during September - December, 2017.

Table 10: Immigration process at Micro Level (See Annexure - 2)

Sl.No.	Number of people Immigrated	No.	Year	Place
1.	Brahmins	1	1963	Iragavaram
2.	Natives	3	2016	Hyderabad, Tallapudi
3.	Other Reserved Category	2	2016	-Do-

Migration for Employment (Out migration and immigration process at the Micro Unit level - Vegeswarapuram).

It is seen from Table 9" that the process of outmigration and immigration in Vegeswarapuram is in practice (Vogue). Further, it is evident from Annexure total number of migrants are to a tune of 100 people and they are away from the village and had permanent residence in different geographic locations, as they have worked in those centres and had settled down due to employment. This process has been started from the year 1950 to todate. Among the outmigrants, Brahmins are out-numbered the entireprocess and the second place goes to Naidus. Padmasali, Rajaka, Christians, Muslims and others took third place in the order ranking and their size. Outmigrants have settled down in the metropolitan city of Hyderabad and Secunderabad. The second important centre and later centres goes to Visakhapatnam, Rajahmundry, Polavaram and Jangareddigudem (Ref. Table. 9) for more details refer to Annexure - 2.

Data relating to immigrants is also presented in Table - 10. The teachers who were transferred from different geographic locations to this place have settled down in Vegeswarapuram village. Almost six persons in our sample establishes that they have been residing from their childhood to senior citizen level. This category could be seen in Table - 10. It is further assessed that migration from rural to urban, from urban to rural is also visible and four people have comeback and have settled down in Vegeswrapuram as they have attachment to agricultural land and buildings. But they eventually spend their lives in a rural-urban environment, as their kids are residing in far way places (i.e.) U.S.A.

7. Cultural Factors

Cultural centres may provide a place for interaction due to cultural meet or cultural events 'cultural centre' makes the people, 'acted) reacted and interacted' and also to renew the friendly relationships. There are 6 temples located in this village and help for 'cultural interaction.

Table 11

Sl.No.	Name of the Temple	Festival falls on the day
1.	Sri Anjaneyaswamy Temple	Vaikunta Sudha Chaturthi (May)
2.	Sri Ramalayam Temple	Srirama Navami Chaitra Masa (April)
3.	Sri Ramlayam Temple (Pettavari Veedhi)	Srirama Navami Chaitra Sukla Navami - (April)
4.	Sri Ramalayam Temple Near Kattula Apparao's House	Srirama Navami Chaitra Sukla Navami (April)
5.	Srimati Tummalamma MaJ'idi Mahalakshmi Temple (Ammavaru)	Vaisakha poornima (May)
6.	Sri Venkateswara Swamy Temple	Kartika Poornima (November)

Source: Author's field work during September - December, 2017.

Sri Kaigala Surya Chandra Trust has been established in the site of I.L.T.D. purchasing centre, Vegeswarapuram. Sri Kaigala Surya Chandra Rao Naidu garu has constructed Sri Venkateswara Swamy temple in 1967 and on Karthikapurnima day, Sri Venkateswara Swamy's festival will be celebrated. On the South-Eastern fringe of the trust land, Sri Kurasala Narasinga Rao and Smt. Varalakshmi have constructed the R.C.C. building for residence. Srikaigala Surya Chandra Rao Trust, Vegeswarapuram has constructed Sri Shiridi Saibaba Temple by Dr. Kaigala Venkata Subba Rao and Smt. Nalini, on 31st July, 2006.

8. Environmental Impact Assessment (EIA) Development At Gross Roots Level

1. Cultural Events: Akhanda Maha Gowthami Pushkaralu, 2015.

During the Godavari Pushkaralu, 2015, C.C. Road has been constructed from Sri Kurasala Satyam Rice Mill to Sri Vinayaka Temple and the length is to tune of 0.475 km, with an estimated cost of Rs.30,00,000 and the funds were made available from C.R.R., during the year, 2014-2015. The work has been started on 25-04-2015 and was successfully been completed by Sri Kalagala Ganeswara Rao, Contractor, Kovvur (Panchayatiraj Department).

2. Mahatma Gandhi National Rural Employment Guarantee Scheme (MGNREGS) was started by Panchayati Raj Department, Government of Andhra Pradesh in Vegeswarapuram, Tallapudi Mandal. C.C. Road has been laid from Srikuncha Adinarayana house to Sri Badara Ganga Rao' s house with an estimated cost of Rs.3.00 lakhs and the funds were made available from MGNREGS 50 per cent and the remaining funds were made available by West Godavari Zilla Parishad, Eluru and the high way stretched to a tune of 95.00 m.

3. From R & B Road to AGRB Ghattam, Sri Venkateswara Swamy Ghattam, Sri Venkateswara Swamy Devalayam the length stretches from 0-83 meters" C.R.R. plain during 2014-2015 and the work was started on 25-4-2015 and successfully completed as per the time tag.

4. Ryutu Sadhikara Samsta, Rytu Runa Upasamana Padhakam is an important scheme has been started by Government of Andhra Pradesh.

Table 12

Sl.No.	Political/Adm-segment	Total amount Rs. In Lakhs Rs.	No. of Rytu Runakatala Sankhya benificiaries
1.	Vegeswarapuram village	25.25	430
2.	Tallapudi Mandalam	26.39 (crores)	3,764
3.	Kovvur (Legislative Assembly segment)	93.25 (crores)	12,626
4.	West Godavari District	1683.45 (crores)	2,53,058
5.	Andhra Pradesh	24,000 (crores)	54.98 (lakhs)

Source: Grama Panchayat Office, Vegeswarapuram (Field work September - December, 2017).

It is seen from Table - 12, that the Rytu Sadhikara Samsta, Rytu Runa Upasamana Padhakam play an important role in agricultural sectoral development and this particular village has been benifitting through this scheme to a tune of Rs.25.25 lakhs and the economic beneficiaries were 430. The Mandal level, Kovvur legislative Assembly segment, West Godavari District and also at the level of Andhra Pradesh state's data is visualized in the line of hierarchie unit development with their total allotment as well as the number of beneficiaries through these schemes.

9. Gross Observations/Changes Over A Period Of Time

1. River water, surface water/under-ground water/soil, vegetation and agriculture should be clubbed in-variably with the socio-economic and infrastructural over heads to understand the environmental impact and its assessment typology.

2. In the areas of agricultural sector, it is concluded that more diversification took place but towards higher value crops, in accordance with the changing water availability, weather conditions access to credit and risk. It is optional, that it goes against the normal feature of crop diversification where farmers diversify in the form of high value crops or the crops which seem to provide higher relative return to the cultivators. Generally, the law of diminishing returns invariably operate, which is an accidental incident which could not be avoided.

3. It is concluded that, analysis of impact of irrigation in terms of gross sown area (GS), cropping intensity (Cl) extent of irrigation (EIR), type of irrigation (TIR) and cropping pattern supports the analysis (1975-2017). However, the net changes have been identified the patterns of changes during the pre and post construction period due to the commencement of new irrigation project in moderate and dry regions that were sponsored by the state.

4. It has been identified that, 'rural development relates economic betterment of people and greater social transformation (ST) - regional development refers to the quality of life at the micro level unit - Vegeswarapuram in terms of economic growth, social development, political maturity and environmental conservation - village development is inclusive of rural development. These factors were analyzed in this paper. Rural is segment of a region, separately it will invite injustice with urban segment of the region.

5. It is observed that the physical factors (land form, soil, rainfall), economic factor, social infrafactors and cultural factor's role is most

important in planning and development. This exercise proved that development takes place when these factors combinedly associate to contribute for development with different development plans over a period of time.

In the study area, the North-West and south East part has been developing so rapidly, the urban process takes a new turn for development in the areas of north of Vegeswarapuram, Tallapudi, Prakhilanka, Gajjaram, Annadevaapaeta, Tallapudi and Vegeswarapurm. These villages have been contributing to a tune of 47.38 per cent.

Population of the southern part of Vegeswrapuram forms as major block with the villages of Ballipadu, Peddavarem Thimgudumetta, Malakapalli and Rvulapadu with a population of 19,025 to a tune of 52.62 per cent.

Tallapudi = 2820 +, Prakkilanka = 3510, Annadevarapeta = 8671, Tallapudi = 4411, 1 Vegeswarapuram = 5601, Total = 25013.

Ballipadu =14011, Peddevam = 6564, Tirugudumetta = 3822, Mallakapalle = 5459, Ravrupadu = 1776, Total = 19025.

It is evident that the southern part of the mandal has higher population but the development is slow, but not natural development. This block forms with a population of 19,025 almost 52.62 per cent.

It has been further identified that the North block forms with a total population of 25,013 which is to a tune of 47.38 percent. Though the populatin is less, the ribbon type of urban development along the highway is on a high pedestal because the mandaI headquarters has also been located geographically vulnerable but not viable.

6. There is every need to regain the P.S. Library With a good building and personal to facilitate people with this facility for continuing academic activity of the people of this village.

7. It is a dire necessity and need to regain the facility of co-operative society i.e, the Primary Agricultural Co-operative Society (PACS), Vegeswarapuram at the earliest possible convenience to cater to the needs of the people, which helps the farmers for carrying out agricultural operations.

8. While providing the administrative infra facilities, it is quite difficult for any governance to provide all the facilities which a mandaI headquarters should possess. It has been further identified that, "Tallapudi Mandal's head quarters is not located in Tallapudi town but it is located in

'Prakkilanka' village, where in the R & B buildings and the vacant sites are also available to provide further infrastructure. Paradoxically, the mandal headquarters of Tallapudi is located in 'Prakkilanka' and its head quarters could not be changed administratively. The primary health centre. (P.H.C.) and police station, besides agricultural extension offices are located exclusively in Prakkilanka.

9. The governance may have to take an active and interactive step by transferring the same from Prakkilanka to Tallapudi. This is an administrative lacuna. If it is changed, then only the mandal development will take place permanently, and there is every need to locate health centre, Government Degree college in Vegeswarapuram village, besides a government of India Bank, for ex-state Bank of India or any other bank to fulfill the needs of the people in this village.

References

1. Krishna Mohan- "Addressing Regional Backwardness: An Analysis of area Development programmes in India" (2005).
2. V. Nath - "Rural Development and Planning in India" (2010)
3. Anita Kumari - "Essentials in Ecology" (2011)
4. Srivastav, O.S. - "Regional Economics and Regional Planning" (2009)
5. Shukla (Amitabh - "Regional Planning and Sustainable Development" (2012)
6. Chauhan, Singh, S.K. -"Environmental Agriculture and Population (P.R) Growth (2014).
7. Ruth, S.K. - "Land use Change and Agricultural Growth in India, (2013).
8. Ferroni, Macro - "Transforming Indian Agriculture India - 2040" (2013)
9. Singh, I - "Integrated Area Development and Planning" (1998)
10. Murty, K, L.N. - "Planning for Integrated Area Development" (1986).
11. Annexure - 1 At case study from Andhra Pradesh
No.1 Land Utilization in Vegeswarapuram
12. Annexure - 2 - The Deccan Geographier, Hyderabad No.2, 1976
13. Annexure - 3 - Geographical Society, Hyderabad, Andhra Pradesh India
(First Survey 1975 & 2nd survey, 2017)
14. Annexure - 4

Annexures

Annexure - I

State	District	Sub distt	Town/Villa	Ward	Level	Name	TRU	*No. of House-hold*	*Total population Person*	*Total population Male*	*Total population Female*	*Population in the age group 0-6 Person*
28	546	00000	000000	0000	DISTRICT	West Godavari	Total	1091525	3936966	1964918	1972048	382887
28	546	00000	000000	0000	DISTRICT	West Godavari	Rural	877207	3128189	1567013	1561176	309636
28	546	00000	000000	0000	DISTRICT	West Godavari	Urban	214318	808777	397905	410872	73251
28	546	04950	000000	0000	SUB-DISTRI	Tallapudi	Total	15109	52785	26295	26490	5612
28	546	04950	000000	0000	SUB-DISTRI	Tallapudi	Rural	15109	52785	26295	26490	5612
28	546	04950	000000	0000	SUB-DISTRI	Tallapudi	Urban	0	0	0	0	0
28	546	04950	588111	0000	VILLAGE	Tallapudi	Rural	568	1896	946	950	199
28	546	04950	588112	0000	VILLAGE	Fagolapalle	Rural	231	802	396	406	77
28	546	04950	588113	0000	VILLAGE	Thupakulagudem	Rural	322	1078	546	532	126
28	546	04950	588114	0000	VILLAGE	Pochavaram	Rural	285	955	463	492	60
28	546	04950	588115	0000	VILLAGE	Paidimetta	Rural	353	1187	577	610	113
28	546	04950	588116	0000	VILLAGE	Prakkilanka	Rural	766	2820	1385	1435	321
28	546	04950	588117	0000	VILLAGE	Gajjaram	Rural	957	3510	1759	1751	380
28	546	04950	588118	0000	VILLAGE	Annadevarapeta	Rural	2549	8671	4325	4346	950
28	546	04950	588120	0000	VILLAGE	Kukunuru	Rural	807	2329	1390	1439	280
28	546	04950	588121	0000	VILLAGE	Thallapudi	Rural	1151	4411	2242	2169	480
28	546	04950	588122	0000	VILLAGE	Vegeswarapuram	Rural	1598	5601	2794	2807	649
28	546	04950	588123	0000	VILLAGE	Ballipadu	Rural	412	1404	697	707	137
28	546	04950	588124	0000	VILLAGE	Peddevam	Rural	1909	6564	3312	3252	657
28	546	04950	588125	0000	VILLAGE	Tirugudumetta	Rural	1110	3822	1865	1957	410
28	546	049:.i0	588127	0000	VILLAGE	Malakapalle	Rural	1569	5459	2706	2753	584
28	546	04950	588128	0000	VILLAGE	Ravurupadu	Rural	522	1776	892	884	189

Population in the age group 0-6 Male	Population in the age group 0-6 female	Scheduled Castes population Person	Scheduled Castes population Male	Scheduled Castes population female	Scheduled Tribes population Person	Scheduled Tribes population Male	Scheduled Tribes population female	Literates population Person	Literates population Male	Literates population female	Illiterate Person	Illiterate Male	Illiterate female
194960	187927	811698	403693	408005	109072	53367	55705	2652389	1379223	1273166	1284577	585695	698882
157580	152056	710703	355154	355549	99659	48661	50998	2035630	1063675	971955	1092559	503338	589221
37380	35871	100995	48539	52456	9413	4706	4707	616759	315548	301211	192018	82357	109661
2852	2760	16951	8439	8512	742	379	363	33505	17427	16078	19280	8868	10412
2852	2760	16951	8439	8512	742	379	363	33505	17427	16078	19280	8868	10412
0	0	0	0	0	0	0	0	0	0	0	0	0	0
95	104	212	109	103	3	2	1	1231	649	582	665	297	368
44	33	372	181	191	0	0	a	510	242	268	292	154	138
62	64	139	73	66	13	7	6	600	310	290	478	236	242
24	36	379	180	199	a	0	a	707	368	339	248	95	153
54	59	475	237	238	1	0	1	767	407	360	420	170	250
163	158	517	268	249	21	9	12	1729	908	821	1091	477	614
193	187	1744	877	867	63	32	31	2236	1194	1042	1274	565	709
495	455	3732	1868	1864	98	52	46	5001	2640	2361	3670	1685	1985
144	136	53	25	28	16	9	7	1987	1018	969	842	372	470
247	233	1136	602	534	170	92	78	2974	1580	1394	1437	662	775
318	331	712	364	348	214	107	107	3570	1826	1744	2031	968	1063
68	69	745	376	369	a	0	0	898	442	456	506	255	251
327	330	2825	1381	1444	83	43	40	4254	2224	2030	'	2310	1088
202	208	1393	675	718	24	9	15	2235	1151	1084	1587	714	873
317	267	2273	1113	1160	30	14	16	3687	1910	1777	1772	/96	976
99	90	244	110	134	6	3	3	1119	558	561	657	334	323

Total Worker Population Person	Total Worker Population Male	Total Worker Population female	Main Working population Person	Main Working population Male	Main Working population Female	Main Cultivator population Person	Main Cultivator population Male	Main Cultivator population Female	Main Agricultural Labourers population Person	Main Agricultural Labourers population Male	Main Agricultural Labourers population Female	Main Household Industries population Person	Main Household Industries population Male
1773112	1199338	573774	1534166	1100258	433908	149420	138120	11300	888815	564273	324542	25703	16859
1485361	974645	510716	1274245	891408	382837	143524	132469	11055	844279	532724	311555	20466	13329
287751	224693	63058	259921	208850	51071	5896	5651	245	44536	31549	12987	5237	3530
24180	16242	7938	23466	15980	7486	2008	1959	49	15720	9516	6204	298	181
24180	16242	7938	23466	15980	7486	2008	1959	49	15720	9516	6204	298	181
0	0	0	0	0	0	0	0	0	0	0	0	0	0
740	590	150	722	577	145	118	117	1	482	357	125	3	2
401	262	139	362	254	108	65	64	1	257	159	98	1	1
573	353	220	567	350	217	38	38	0	433	244	189	5	3
415	299	116	353	277	76	110	107	3	201	131	70	0	0
542	361	181	532	359	173	42	41	1	282	156	126	1	1
1281	827	454	1235	814	421	22	22	0	507	233	274	34	25
1706	1088	618	1690	1083	607	159	157	2	1207	680	527	6	4
4675	2734	1941	4554	2709	1845	321	307	14	3739	2031	1708	12	10
1031	838	193	1014	830	184	62	62	0	372	264	108	11	7
1748	1294	454	1738	1286	452	31	30	1	585	381	204	20	12
2332	1727	605	2082	1628	454	124	121	3	1145	839	306	62	46
694	466	228	682	460	222	127	127	0	437	254	183	22	11
2861	2026	835	2829	2007	822	192	182	10	2284	1551	733	7	7
1932	1189	743	1316	1181	735	203	197	6	1574	885	689	14	9
2465	1597	868	2415	1579	836	274	268	6	1606	907	699	87	38
784	591	193	775	586	189	120	119	1	609	444	165	13	5

Main Household Industries population Female	Main Other Worker Population Person	Main Other Worker Population Male	Main Other Worker population Female	Marginal Worker Population Person	Marginal Worker Population Male	Marginal Worker population Female	Marginal Cultivator Population Person	Marginal Cultivator Population Male	Marginal Cultivator population Female	Marginal Agriculture LabourersPopulation Person	Marginal Agriculture LabourersPopulation Male	Marginal Agriculture LabourersPopulation Female	Marginal Household Industries Population Person
8844	470228	381006	89222	238946	99080	139866	5855	3283	2572	172972	67484	105488	8158
7137	265976	212886	53090	211116	83237	127879	5580	3091	2489	163991	62816	101175	6851
1707	204252	168120	36132	27830	15843	11987	275	192	83	8981	4668	4313	1307
117	5440	4324	1116	714	262	452	21	14	7	372	142	230	63
117	5440	4324	1116	714	262	452	21	14	7	372	142	230	63
0	0	0	0	0	0	0	0	0	0	0	0	0	0
1	119	101	18	18	13	5	0	0	0	11	6	5	5
0	39	30	9	39	.	8	31	3	3	0	31	5	26
2	91	65	26	6	3	3	1	1	0	3	1	2	0
0	42	39	3	62	22	40	1	1	0	42	16	26	3
0	207	161	46	10	2	8	0	0	0	1	0	1	0
9	672	534	138	46	13	33	1	1	0	25	3	22	3
2	318	242	76	16	5	11	0	0	0	8	1	7	0
2	482	361	121	121	25	96	1	0	1	82	18	64	4
4	569	497	72	17	8	9	4	4	0	4	1	3	0
8	1102	863	239	10	8	2	2	1	1	0	0	0	0
16	751	622	129	250	99	151	3	1	2	111	68	43	41
11	96	68	28	12	6	6	0	0	0	2	1	1	2
0	346	267	79	32	19	13	5	2	3	16	9	7	1
5	125	90	35	16	8	8	0	0	0	5	2	3	2
49	448	366	82	50	18	32	0	0	0	22	6	16	0
8	33	18	15	9	5	4	0	0	0	9	5	4	0

Marginal Household Industries Population Male	Marginal Household Industries Population Female	Marginal Other Workers population Person	Marginal Other Workers population Male	Marginal Other Workers population Female	Marginal Worker population 3-6 Person	Marginal Worker population 3-6 Male	Marginal Worker population 3-6 Female	Marginal Cultivator population 3-6 Person	Marginal Cultivator population 3-6 Male	Marginal Cultivator population 3-6 Female	Marginal Agriculture Labourer Population Person	Marginal Agriculture Labourer Population Male	Marginal Agriculture Labourer Population Female
2125	6033	51961	26188	25773	198892	86481	112411	5089	2956	2133	145110	59207	85903
1537	5314	34694	15793	18901	175918	73075	102843	4825	2772	2053	137844	55367	82477
588	719	17267	10395	6872	22974	13406	9568	264	184	80	7266	3840	3426
12	51	258	94	164	587	222	365	21	14	7	304	125	179
12	51	258	94	164	587	222	365	21	14	7	304	125	179
0	0	0	0	0	0	0	0	0	0	0	0	0	0
5	0	2	2	0	9	5	4	0	0	0	8	4	4
0	2	3	0	3	32	6	26	3	3	0	26	3	23
0	0	2	1	1	6	3	3	1	1	0	3	1	2
0	3	16	5	11	62	22	40	1	1	0	42	16	26
0	0	9	2	7	8	2	6	0	0	0	1	0	1
2	1	17	7	10	37	11	26	1	1	0	20	3	17
0	0	8	4	4	8	4	4	0	0	0	3	1	2
2	2	34	5	29	81	18	63	1	0	1	52	11	41
0	0	9	3	6	14	8	6	4	4	0	4	1	3
0	0	8	7	1	4	3	1	2	1	1	0	0	0
1	40	95	29	66	237	93	144	3	1	2	100	63	37
1	1	8	4	4	9	5	4	0	0	0	2	1	1
0	1	10	8	2	30	17	13	5	2	3	16	9	7
1	1	9	5	4	12	6	6	0	0	0	3	2	1
0	0	28	12	16	30	14	16	0	0	0	16	5	11
0	0	0	0	0	8	5	3	0	0	0	8	5	3

Marginal Household Industries Population 3-6 Person	Marginal Household Industries Population 3-6 Male	Marginal Household Industries Population 3-6 Female	Marginal Other Workers Population Persons 3-6 Person	Marginal Other Workers Population Persons 3-6 Male	Marginal Other Workers Population Persons 3-6 Female	Marginal Workers Population 3-6 Person	Marginal Workers Population 3-6 Male	Marginal Workers Population 3-6 Female	Marginal Cultivator Population 3-6 Person	Marginal Cultivator Population 3-6 Male	Marginal Cultivator Population 3-6 Female	Marginal Agriculture Labourers Population 3-6 Person	Marginal Agriculture Labourers Population 3-6 Male
6179	1700	4479	42514	22618	19896	40054	12599	27455	766	327	439	27862	8277
5223	1261	3962	28026	13675	14351	35198	10162	25036	755	319	436	26147	7449
956	439	517	14488	8943	5545	4856	2437	2419	11	8	3	1715	828
55	6	49	207	77	130	127	40	87	0	0	0	68	17
55	6	49	207	77	130	127	40	87	0	0	0	68	17
0	0	0	0	0	0	0	0	0	0	0	0	0	0
0	0	0	1	1	0	9	8	1	0	0	0	3	2
0	0	0	3	0	3	7	2	5	0	0	0	5	2
0	0	0	2	1	1	0	0	0	0	0	0	0	0
3	0	3	16	5	11	0	0	0	0	0	0	0	0
0	0	0	7	2	5	2	0	2	0	0	0	0	0
3	2	1	13	5	8	9	2	7	0	0	0	5	0
0	0	0	5	3	2	8	1	7	0	0	0	5	0
4	2	2	24	5	19	40	7	33	0	0	0	30	7
0	0	0	6	3	3	3	0	3	0	0	0	0	0
0	0	0	2	2	0	6	5	1	0	0	0	0	0
41	1	40	93	28	65	13	6	7	0	0	0	11	5
2	1	1	5	3	2	3	1	2	0	0	0	0	0
1	0	1	8	6	2	2	2	0	0	0	0	0	0
1	0	1	8	4	4	4	2	2	0	0	0	2	0
0	0	0	14	9	5	20	4	16	0	0	0	6	1
0	0	0	0	0	0	1	0	1	0	0	0	1	0

Marginal Agriculture Labourers Population 0-3 Male	Marginal Household Industries Population 0-3 Person	Marginal Household Industries Population 0-3 Male	Marginal Household Industries Population 0-3 Female	Marginal Other Workers Population 0-3 Person	Marginal Other Workers Population 0-3 Male	Marginal Other Workers Population 0-3 Female	Non Working Population Person	Non Working Population Male	Non Working Population Female
19585	1979	425	1554	9447	3570	5877	2163854	765580	1398274
18698	1628	276	1352	6668	2118	4550	1642828	592368	1050460
887	351	149	202	2779	1452	1327	521026	173212	347814
51	8	6	2	51	17	34	28605	10053	18552
51	8	6	2	51	17	34	28605	10053	18552
0	0	0	0	0	0	0	0	0	0
1	5	5	0	1	1	0	1156	356	800
3	2	0	2	0	0	0	401	134	267
0	0	0	0	0	0	0	505	193	312
0	0	0	0	0	0	0	540	164	376
0	0	0	0	2	0	2	645	216	429
5	0	0	0	4	2	2	1539	558	981
5	0	0	0	3	1	2	1804	671	1133
23	0	0	0	10	0	10	3996	1591	2405
0	0	0	0	3	0	3	1798	552	1246
0	0	0	0	6	5	1	2663	948	1715
6	0	0	0	2	1	1	3269	1067	2202
0	0	0	0	3	1	2	710	231	479
0	0	0	0	2	2	0	3703	1286	2417
2	1	1	0	1	1	0	1890	676	1214
5	0	0	0	14	3	11	2994	1109	1885
1	0	0	0	0	0	0	992	301	691

Annexure - 2

List of Migrants from the Micro level unit, Vegeswarapuram

From one caste put together

	Name	Year - Place
1.	Kalluri Veerappa Somayajulu	1950 - Undi, West Godavari Dist.
2.	Kalluri Ramabhdrachaynulu	1977 - Kovvur, West Godavari Dist.
3.	Akella Sesha Rao	1984 - Hyderabad.
4.	Yeluripati SarvotomukhfYajulu	1950 - Kothapet, East Godavari Dist.
5.	Yeluripati Venkata Sayanarayana Murthy	1985 - Hyderabad.
6.	Nori Narasimha Sastry	1969 - expired
7.	Nori Venkata Subrahmanayam M.E.O. (Retd)	2016 - Hyderabad
8.	Nori Subrahmanayam Postal Dept	1962 - Rajahmundry
9.	Nori Balakrishna	1972 - APSEB, Hyderabad
10.	Nori Suryanarayana Murthy	1974 - IDPL, Hyderabad.
11.	Nori Peri Sastry	1982 - Zilla Parishad School, East Godavari Dist, Kakinada.
12.	Sri Kandukuri Gandhi, AMIE	1950 - Irrigation Department,Nagarjuna Sagar
13.	Sri T.V. Subba Rao, HM (Retd.)	2000 - Gopavaram, Near Nidavaolu,West Godavari Dist.
14.	Sri T.S.R. Murty, Lecturer	1976 - Rajahmundry, East Godavari Dist.
15.	Sri T.G.K. Murty, Teacher	1980 - Z.P. School, East Godavari Dist
16.	Sri K.V. Hanumantha Rao,SBI	1964 - Hyderabad.
17.	Sri K.S.R. Murty	1976 - Narsipatnam,
18.	Sri K.S.N. Murty, Agri.Dept.	1980 - A.P.Sugars
19.	Shaik Masthan, Railway Dept.	1968 - Hyderabad

20. Prof. K.V.Subba Rao, End.IIT,Mumbai- 1967 - Canada

21. Sri Angerapu Tumma Raju, Edn. Dept.- 1967 - Polavaram, West Godavari Dist

22.	Sri AngarapuTata Rao, Edn.Dept.	1987 - Jangareddygudem, West Godavari Dist
23.	Sri Gullala Sita	1975 - Visakhapantam
24.	Sri Kanuri Atchanna, Edn.Dept.	1962 - Z.P. Eluru, West Godavari Dist
25.	Sri K.S.P. Murthy	1968 - Hyderabad
26.	Sri K.C.S. Murty	1971 - Visakhapatnam
27.	Sri K.L. Narasimha Murthy	1976 - Rajahmundry
28.	Sri K.R.C. Murthy	1977 - Hyderabad
29.	Sri K.S.K. Murthy	1983 - Hyderabad
30.	Sri K.S.D. Murthy	1983 - Pulla/Hyderabad/Rajahmundry
31.	Sri K.V. Somayajulu	1982 - Rajahmundry/Hyderabad
32.	Dr.A.V.B. Sankaram, RR Labs	1960 - Andhra Univeri sty, Visakhapatnam And Hyderabad.
33.	Dr. A. Purushottam, OSC	1950 - Andhra University, Visakhapatnam and Kolkota.
34.	Sri A.V.S.R. Charyulu, MSC	1960 - GSI, Hyderabad
35.	Sri Someswara Sarma, BSC.	1960 - Madanapalle, Chittoor Dist.
36.	Sri A. Viswanadham	1966 - GSI, Hyderabad
37.	Dr. G. Suryanarayana, Ph.D.	1976 - Visakhapatnam / Machilipatnam
38.	Sri Yeluripati Chaynulu	1964 - Hyderabad
39.	Sri Y.V.A.V. Bhaskhara Rao	1975 - Hyderabad
40.	Miss. Y. Bhaskar Lakshmi	1980 - Hyderabad
41.	Sri Y. Babu	1982 - Hyderabad
42.	Miss. Y. Prabhavathi	1983 - Hyderabad
43.	Miss. Y. Srivalli	1985 - Hyderabad
44.	Sri Chelukuri Ramabhadra Sastry	1977 - Tanuku
45.	Sri G.V. Satyanarayana Murty	1970 - Kovvur
46.	Sri G.V. Satyanarayana Murty	1980 - Kovvur
47.	Sri G. Babi	1964 - Nagpur
48.	Sri M. Sriram Sarma	1959 - Hyderabad
49.	Sri M. Ramakrishna Rao	1964 - Hyderabad
50.	Sri M. Suryanarayana Murthy	1966 - Hyderabad
51.	Sri Konala Srirama Murthy	1960 - Rajahmundry
52.	Sri Konala Satyanarayana, RBI	1960 - Hyderabad
53.	Sri Addala Tata Rao	1961 - Hyderabad
54.	Sri K.S.R.Ch.V.K. Chinta Rao, APCOBank	1971 - Hyderabad
55.	Sri V. Ganeswara Rao	1971 - Hyderabad

56.	Sri K. Padma Raju	1965 - Hyderabad
57.	Sri Somanchi Narasimha Murty	1962 - Hyderabad
58.	Sri Ch. Krishna Murthy	1970 - Kakinada
59.	Sri Ch. Viswanadha Sastry	1965 - Nagarjuna Sagar
60.	Sri Ch. Veerabhadra Rao	1980 - Nagarjuna Sagar
61.	Sri Ch. Sitaiah	1960 - Rajahmundry
62.	Sri Ch. Rambabu	1961 - Buttaigudem
63.	Ch. Venugopala Rao	1970 - Dharmavaram
64.	Sri Guttikonda Satyanarayana	1971 - Hyderabad
65.	Sri T.V.V. Satyanarayana	1966 - Kovvur (AS Ltd.)
66.	Sri P. Lakshmana Rao	1976 - Telephone Exchange, Rangareddi gudem
67.	Sri Ch. Lakshmipati	1980 - Kakinada
68.	Sri D. Jagan Mohan Rao	1965 - Eluru
69.	Sri D.V. Suresh Chandraji	1978 - Bhimavaram
70.	Sri D. Prasad	1985 - U.S.
71.	Sri Talla Bangarraju	2000 - Visakhapatnam
72.	Sri Chamarti Satyanarayana M.A., B.Ed. (MEO) (Retd.)	1962 - Rajahmundry
73.	Sri Sripada Harinath	1960 - Rajahmundry

Annexure - 3

Never Moved from Native place

1.	Sri K. Satyanarayana	1940
2.	Sri Madhavarapu Venkatea Rao	1944
3.	Sri Konala Narayana Rao	1947
4.	Sri A. Venkateswara Rao	1944
5.	Sri K. Venkateswara Rao	1940
6.	Sri K.V. Chalam	1940

Annexure - 4

In Migration as on date 1-11-2017

1.	Sri Akella Kama Sastry M.A., B.Ed., H.M (Retd.)	1963 - Vegeswarapuram
2.	Sri N. Suryanarayana Murthy M.Pharma, (Retd.)	2017 - From Hyderabad
3.	Sri N. Peri Sastry M.A., M.Ed., (Retd.)	2017 - From East Godavari, West Godavari
4.	Sri D.V. Narasimha Murthy (Retd)	1950 - K.S.R.Z.P.H. School, Vegeswarapuram
5.	Smt. Prasadamma, Teacher (Retd)	1956 - P.S.J.B. School, Vegeswarapuram
6.	Sri Krupavaram (Retd.) Teacher	1970 - P.S.J.B. School, Vegeswarapuram
7.	Smt. Samuel (Retd.) Teacher	1955 - P.S.J.B. School, Vegeswarapuram
8.	Sri K.S.R.Ch.V.K.Chinta Rao (Retd)	2010 - A.P. Co-operative Services, Hyderabad
9.	Sri V. Satyanarayana (Doctor)	2005 - Educational Dept., Vegeswarapuram
10.	Sri T.G.K. Murty (Homoeo Doctor)	2014 - Educational Dept., Vegeswarapuram
11.	Sri Dhulipala Venkateswarlu	1963 - Agricultural Migration (Operation)
12.	Sri M.V. Satyanarayana	2008 - ITDA, Visakhapatnam to V.G.P. Butaigudem

Annexure - 5

G.I.S. of Vegeswarapuram

Village	Vegeswarapuram
Block	Thallapudi
District	West Godavari
State	Andhra Prdesh
Country	India
Continent	Asia
Time Zone	IST (UTC + 05:30)
Currency	Indian Rupee (INR)
Dialing Code	+91
Date fromat	dd/mm/yyyy
Driving side	left
Internet cTLD	in
Langunage	Telugu, Urdu, English
Time difference	16 minutes
Latitude	17.4
Longitude	78.48

Vegeswarapuram is a village panchayat located in the West Godavari district of Andhra-Pradesh state ,India. The latitude 17.4 and longitude 78.48 are the geocoordinate of the Vegeswarapuram.

The official language of Vegeswarapuram

The native language of Vegeswarapuram is Telugu , Urdu, English and most of the village people speak Telugu, Urdu, English. Vegeswarapuram people use Telugu, Urdu, English language for communication.

Vegeswarapuram Sun rise time

Vegeswarapuram village is located in the UTe 5.30 time zone and it follows indian standard time(IST). Vegeswarapuram sun rise time varies 16 minutes from 1ST. The vehicle driving side in Vegeswarapuram is left, all vehicles should take left side during driving. Vegeswarapuram people are using its national currency which is Indian Rupee and its internationl currency code is INR. Vegeswarapuram phones and mobiles can be accesed by adding the Indian Country Dialing Code +91 from abroad. Vegeswarapuram people are following the dd/mm/yyyy date format in day-to-day life. Vegeswarapuram domain name extension (cTLD) is .in.

The nearest railway station in and around Vegeswarapuram – Kovvur.

Nearest airport to Vegeswarapuram – Rajahmundry

Nearest districts to Vegeswarapuram – Eest Godavari

Nearest town/city to Vegeswarapurarn – Nidadevole and Tadipalli-Gudem

Schools in and around Vegeswarapuram – Malakapalli; Tallapudi; Annadevaripeta; Gutala

Index

O

P

Q

R

T

U

V

W

Y

Z